Genetic Engineering 5

Genetic Engineering 5

Edited by

Peter W. J. Rigby

Cancer Research Campaign,
Eukaryotic Molecular Genetics Research Group,
Department of Biochemistry,
Imperial College of Science and Technology,
London

ACADEMIC PRESS · 1986

Harcourt Brace Jovanovich, Publishers
London · Orlando
New York · San Diego · Austin
Montreal · Sydney · Tokyo · Toronto

Printed in Great Britain at the Alden Press
Oxford London and Northampton

ACADEMIC PRESS INC. (LONDON) LTD
24/28 Oval Road,
London NW1

United States Edition published by
ACADEMIC PRESS INC.
Orlando, Florida 32887

British Library Cataloguing in Publication Data

Genetic engineering
  5
  1. Genetic engineering—Periodicals
  I. Rigby, Peter
  575.1   QH442
  ISBN 0-12-270305-7

Contributors

A. Hall *Chester Beatty Laboratories, Institute of Cancer Research, Fulham Road, London SW3 6JB, UK*

C.F. Higgins *Department of Biochemistry, University of Dundee, Dundee DD1 4HN, UK*

M. Steinmetz *Basel Institute for Immunology, Grenzacherstrasse 847, CH-4005, Basel, Switzerland*

Preface

In 1981, when Bob Williamson edited the first volume in this series, genetic engineering was a sophisticated activity practised only in specialized research laboratories and a few biotechnology companies. In the past five years it has expanded at a rate which even the most optimistic of its pioneers could only have guessed at. It now plays a central role in almost all aspects of biology and has become established as a tool in industry. Genetic engineering is not a subject as such but a technology of extraordinary power which will underpin much of biological research, whether pure or applied, for the forseeable future.

Such change necessitates corresponding change in the style of this series. The first four issues concentrated on the technology which has to be mastered by all who wished to do this type of work. Much of that material is now covered in undergraduate courses and will be assumed in this volume and its successors. Moreover, there are now available several excellent laboratory manuals which detail the relevant experimental procedures. In the four volumes which I shall edit the results obtainable using the methods of genetic engineering and the concepts to which they have led will be emphasized. However, technical improvements continue apace and will be reviewed where appropriate.

This volume contains three articles in areas where genetic engineering has made major contributions to knowledge. The initial excitement engendered by our ability to clone and characterize genes stemmed almost exclusively from the application of the techniques to eukaryotic genes. It was too often forgotten that the methods are equally applicable to the study of gene regulation in prokaryotes. This area has recently undergone a renaissance. Chris Higgins discusses the important new advances and describes some new technologies. The articles which follow deal with the two areas of eukaryotic molecular biology which perhaps have been most dramatically changed by the application of gene manipulation techniques. Alan Hall's chapter deals with one of my own research interests, the molecular biology of cancer. I doubt that anyone could have predicted the extraordinary progress which has brought not only a fundamental change in our understanding of the problem but also the prospect of improved diagnostic procedures and the possibility of new modes of treatment. The generation of immunological diversity used to be a major theoretical problem for biologists, yet we now have a precise picture of how it is achieved and we are beginning to understand some of the mechanisms involved. Michael Steinmetz reviews our knowledge of the genes of the immune system, those encoding immunoglobulins, the T cell antigen receptor and the molecules of the major histocompatibility complex, providing an excellent introduction to this rapidly moving field.

In his preface to Volume 4, my predecessor commented that one major omission from the series was a discussion of the application of genetic engineering techniques to plants. This will be the topic of Volume 6. Subsequent volumes will include articles on site-directed mutagenesis and protein engineering, molecular parasitology, genes involved in embryonic development and the genetic manipulation of eukaryotic cells and organisms.

The first four volumes in this series were very well received by their readers. It is my hope that the four for which I am responsible will be as successful in stimulating interest in the research and in helping teachers and students. I have been set a high standard.

I should like to thank the authors for the hard work and expertise that they have brought to this volume. I am grateful to Bob Williamson and the staff of Academic Press who persuaded me to undertake this task, and to the present staff of the Press for their encouragement and forbearance.

London, February 1986   *Peter W. J. Rigby*

Contents

Contributors   v

Preface   vii

The regulation of gene expression in *Escherichia coli* and its bacteriophage

C. F. Higgins

I	Introduction	2
II	Regulation of transcription	3
	A Initiation of transcription	3
	B Elongation of mRNA	12
	C Termination of transcription	13
III	Regulation of RNA processing and degradation	21
	A mRNA secondary structure	22
	B Effects of protein binding on mRNA degradation	23
	C Endonuclease cleavage of mRNA	24
IV	Regulation of DNA rearrangement	28
	A Random rearrangements	28
	B Controlled rearrangements.	30
V	Translational regulation	32
	A Signals encoded within the mRNA.	33
	B Regulation by masking of ribosome-binding sites	36
VI	Gene and operon fusions	41
	A Construction of gene and operon fusions *in vivo*	43
	B Construction of gene and operon fusions *in vitro*	48
	C Uses of fusion technology	53
VII	Conclusions	54
VIII	Acknowledgements.	55
IX	References.	55

Oncogenes

Alan Hall

I	Introduction	61
II	The transformed cell	63
	A Density-dependent inhibition of growth	65
	B Anchorage-independent growth	65
	C Immortality.	67
	D Growth factor requirements	68
	E *In vivo* tumorigenicity	69
	F Metastasis	69
III	RNA tumour viruses and viral oncogenes	69
	A Rous sarcoma virus.	70
	B Other avian viruses	72
	C Mammalian viruses.	74
IV	Cellular oncogenes.	75
	A Proto-oncogenes	75
	B Proto-oncogenes and viral oncogenes	76
	C Viral activation of proto-oncogenes	79
	D Non-viral activation of proto-oncogenes	83
	E Biological assays to detect cellular oncogenes.	92
V	Biochemistry of oncogene proteins	99
VI	Summary	108
VII	Acknowledgements.	109
VIII	References	109

Genes of the immune system

Michael Steinmetz

I	Introduction	118
II	Major histocompatibility complex	119
	A Structure and function of major histocompatibility complex class I and class II molecules	119
	B Genetic map	121
III	Immunoglobulin genes	123
	A Structure of immunoglobulin molecules	123
	B The Dreyer–Bennett hypothesis	124
	C Organization and rearrangement of immunoglobulin genes	125
IV	T-cell receptor genes	129
	A The T-cell receptor complex	129

B   Organization and rearrangement of T-cell receptor α and
    β chain genes 130
C   The T-cell-receptor related γ chain gene 132
V   The immunoglobulin superfamily 133
A   Sequence relationships 133
B   The immunoglobulin homology unit 134
C   Rearranging immunoglobulin and T-cell receptor gene
    segments 134
D   Chromosomal locations and translocations . . . 136
VI Generation of diversity 138
A   Major histocompatibility complex alleles 138
B   Immunoglobulin and T-cell receptor genes 141
VII Species comparison 145
A   Major histocompatibility complex genes 145
B   Immunoglobulin genes 146
C   T-cell receptor genes 147
VIII Conclusions 147
IX Acknowledgements 148
X   References 148

The regulation of gene expression in *Escherichia coli* and its bacteriophage

C. F. HIGGINS

Department of Biochemistry, University of Dundee, Dundee DD1 4HN, UK

I	Introduction	2
II	Regulation of transcription	3
	A Initiation of transcription	3
	B Elongation of mRNA	12
	C Termination of transcription	13
III	Regulation of RNA processing and degradation	21
	A mRNA secondary structure	22
	B Effects of protein binding on mRNA degradation	23
	C Endonuclease cleavage of mRNA	24
IV	Regulation of DNA rearrangement	28
	A Random rearrangements	28
	B Controlled rearrangements	30
V	Translational regulation	32
	A Signals encoded within the mRNA	33
	B Regulation by masking of ribosome-binding sites	36
VI	Gene and operon fusions	41
	A Construction of gene and operon fusions *in vivo*	43
	B Construction of gene and operon fusions *in vitro*	48
	C Uses of fusion technology	53
VII	Conclusions	54
VIII	Acknowledgements	55
IX	References	55

GENETIC ENGINEERING Vol. 5
ISBN 0-12-270305-7

I Introduction

This year marks the twenty-fifth anniversary of Jacob and Monod's remarkable series of experiments which provided the framework on which most of our current ideas about the regulation of gene expression are based. It therefore seems appropriate to assess our current understanding of the subject and, in particular, the recent renaissance in the study of prokaryotic gene expression which has stemmed from the development of recombinant DNA technology. Not only has the range of genes and organisms which can be studied been dramatically increased but a whole host of new and sophisticated regulatory mechanisms have been uncovered which can now be understood in considerable detail at a molecular level. The standard textbook view, that an understanding of the regulation of the lactose operon and the mechanism of attenuation in the tryptophan operon is all that is required to appreciate the subtleties of bacterial gene expression, is rapidly being demolished. It is becoming clear not only that prokaryotes regulate gene expression at every conceivable level but also that they adopt very subtle fine-tuning mechanisms to adapt the precise expression of any given gene to each particular set of growth conditions.

This chapter will deal almost entirely with *Escherichia coli* and its bacteriophage. This is partly personal prejudice, partly for reasons of space but primarily because *E. coli* is by far the best understood living organism. Most of the ideas and principles behind our understanding of gene expression were developed using *E. coli* as a model system and are best illustrated by examples from this species. However, it is obvious that the regulation of gene expression in other bacterial species can be equally, if not more, subtle and in some cases (e.g. the discovery of multiple σ factors in *Bacillus*) these species have upstaged *E. coli*. It is also worth pointing out that, although little is known about the molecular events which regulate eukaryotic gene expression, most of the models which are being developed are based on concepts derived from prokaryotic studies. After all, it was Monod himself who is reputed to have said, "What is true for *E. coli* is also true for an elephant".

Other chapters in this series have provided excellent descriptions of many of the techniques of cloning and genetic manipulation which have enabled the mechanisms of gene expression to be unravelled. Thus I do not propose to dwell on experimental techniques but on the results which they have provided. The one exception is the construction and use of gene and operon fusions which have revolutionized the study of gene expression. The range of genes which can be readily studied, and the variety of species which are accessible to manipulation, are increasing exponentially with the development of these technologies. A description of gene and operon fusion technology is provided in section VI.

This chapter is intended to provide an overview of our current understanding of the multiple mechanisms by which gene expression can be regulated. In particular, I hope to draw attention to the many different levels of control which can be adopted. Gene expression is regulated not only by influences on transcription initiation but also by the pausing and termination of RNA polymerase, by RNA processing and degradation, by DNA rearrangements and by several ingenious methods of modifying the translational efficiency of mRNA. This chapter is subdivided in order to reflect these various categories of control mechanism. Undoubtedly, other unexpected mechanisms will emerge in the next few years. Unfortunately, space does not permit a detailed discussion of all aspects of bacterial gene expression and several important areas have had to be considered in less detail than they deserve. However, I hope that the intriguing and novel control mechanisms that I have included will serve to illustrate those areas of study in which exciting developments are currently being made.

II   Regulation of transcription

Because bacterial cells must be able to adapt to constantly changing conditions they have evolved the ability to adjust, often extremely rapidly, the genes which are expressed at any given time. The most economical point at which to regulate expression of a gene is generally the first step—transcription. This prevents resources being wasted in the synthesis of unwanted mRNA and has certainly proved to be the major point at which gene expression is controlled. However, transcription is a multistep process which can be conveniently divided into initiation, elongation and termination events. While the most familiar mechanisms for controlling gene expression, such as repressor or activator binding, affect transcription initiation, it is becoming ever more clear that control can also be exerted during elongation and termination.

A   Initiation of transcription

Transcription initiation depends on the correct interaction of RNA polymerase with a promoter sequence on the DNA template. Anything which influences this process provides a means by which gene expression can be controlled. Those factors which affect promoter function can broadly be divided into two classes: (i) features intrinsic to the promoter itself and (ii) external factors which interact with the DNA, or with RNA polymerase, and alter the interaction between the two. An excellent review of protein–DNA interactions has recently been published (von Hippel *et al.*, 1984).

1   *Features intrinsic to the promoter*

The strength of a promoter, and thus the level of expression from it, clearly depends on its sequence. Operationally, the bests means of defining promoters is by combining genetic analysis, the selecting and sequencing of promoter mutations (see for example Beckwith, 1981, Higgins and Ames, 1982, and Youderian *et al.*, 1982), with biochemical analyses such as mapping of the 5′ end of the *in vivo* mRNA and sequencing of the upstream region of DNA. Comparison of promoter sequences from a large number of genes (168 in the latest compilation (Hawley and McClure, 1983)) defines conserved nucleotides at about 10 and 35 base pairs (bp) upstream from the transcription initiation point. Thus a consensus promoter sequence can be derived (Fig. 1). Over 100 promoter mutations have now been sequenced: nearly all those which increase gene expression increase homology with the consensus promoter sequence and nearly all those which decrease expression reduce homology (Youderian *et al.*, 1982; Hawley and McClure, 1983). In addition to the importance of the nucleotide sequence of the − 10 and − 35 regions the spacing between these two sequence blocks and the nucleotide composition of the surrounding regions also influence promoter strength. For example, AT-rich regions, which facilitate "melting" of the DNA double helix, facilitate promoter activity. Sequence differences in and around promoters can also lead to

```
                 -35                      -10        +1
        a......tcTTGACat..t.........t.tg.TAtAat......cat
```

Figure 1   Consensus *E. coli* promoter sequence: the nucleotide most frequently present at each position of a promoter sequence is taken from the analysis of Hawley and McClure (1983). Nucleotides which occur in at least 39% of all promoters (three standard deviations above random occurrence) are listed; those which occur in more than 54% of promoters (six standard deviations above random) are capitalized.

structural polymorphisms, recognized experimentally by altered sus-
ceptibilities to nucleases such as DNaseI or DNaseII, which can affect
promoter activity (Drew and Travers, 1984). It is not yet possible to
predict the relative strength of a promoter, or even to identify a
promoter with certainty, on the basis of nucleotide sequence data alone.
This implies that factors other than the primary sequence of the
promoter itself may play a role in determining function. In eukaryotic
cells enhancer sequences present at some distance from the promoter
can dramatically increase its efficiency. While there is no evidence for
enhancer-like sequences in *E. coli* there is evidence that sequences
upstream of a promoter can influence its function. The most notable
example is the *tyrT* promoter, which is one of the strongest bacterial
promoters. Reverse genetic analysis, involving defined deletion of
sequences upstream from the *tyrT* promoter, shows that, at least in this
case, sequences up to 200bp 5′ to the point of transcription initiation can
profoundly affect promoter function (Lamond ar.d Travers, 1983;
Travers *et al.*, 1983). Although the role of these upstream sequences
remains obscure, DNA footprinting studies suggest that they may serve
as additional, but non-productive, binding sites for RNA polymerase.

A further factor which can influence promoter function is DNA
supercoiling. There is accumulating evidence that the *E. coli* chromosome is divided into many domains, each of which is independently
supercoiled (Sinden and Pettijohn, 1981). Phage and plasmid DNAs are
also supercoiled within the cell. The level of supercoiling is maintained
by a careful balance between the opposing activities of topoisomerase I
and DNA gyrase (DiNardo *et al.*, 1982; Richardson *et al.*, 1984) and it is
clear that the expression of many genes is dependent on the level of
DNA supercoiling (see for example Sternglanz *et al.*, 1981). The clearest
example is the *leu-500* mutation which introduces a single AT-to-GC
base pair change in the −10 region of the leucine operon promoter
(Gemmill *et al.*, 1984). The *leu-500* promoter does not function in wild-
type cells but activity is restored by mutations in the gene for
topoisomerase I, presumably as a result of oversupercoiling of the DNA
template (Mukai and Margolin, 1963; Sternglanz *et al.*, 1981; Richardson
et al., 1984). Topoisomerase and gyrase mutations, or inhibitors of these
enzymes, alter the expression of many but not all genes. In addition, it is
well known that, in *in vitro* transcription systems, some genes are
transcribed more efficiently from linear templates while for other genes
a supercoiled template is preferred. Insertion sequences may also alter
gene expression by altering DNA topology (DiNardo *et al.*, 1982; Stokes
and Hall, 1984). However, there is as yet no conclusive evidence which
shows that DNA supercoiling varies with growth conditions or at

different stages in the cell cycle. Thus the question whether or not DNA supercoiling plays a role in modulating gene expression under different environmental conditions remains open.

One further factor which might influence gene expression is DNA methylation. While the role of methylation has been keenly debated in eukaryotes, little attention has been paid to any possible regulatory role in prokaryotes. Recently, however, it has been clearly demonstrated that methylation of GATC sequences by the host *dam* methylation system is essential for transcription of the *mom* gene of bacteriophage Mu. The mechanism by which methylation stimulates transcription is unknown but resemblances of the *mom* promoter to the origin of replication suggest it may be at the level of DNA structure (Plasterk *et al.*, 1984).

2 Proteins which alter DNA–RNA polymerase interactions: σ factors

Proteins can control the interaction of RNA polymerase with promoters in one of two ways: either they can interact with RNA polymerase and alter its specificity (the σ factors) or they can bind directly to the DNA (positive and negative regulatory proteins).

RNA polymerase consists of five subunits. The core enzyme has the subunit composition $\alpha_2\beta\beta'$ and, in addition, there is a dissociable subunit called σ. σ confers specificity to the interaction between RNA polymerase and the promoter. The possibility that alternative σ factors might direct RNA polymerase to transcribe different sets of genes was first discovered during studies of *Bacillus subtilis* sporulation and the development of the *Bacillus*-specific phage SPO1 (Losick and Pero, 1981). Under stress conditions *Bacillus* undergoes a well-defined differentiation to form resistant endospores. This sporulation process requires the proper sequential expression of a large number of specific genes (the *spo* genes). Several forms of RNA polymerase can be isolated from *Bacillus* cells, at least one of which appears only at the onset of sporulation. These various forms of RNA polymerase have identical core polypeptides and vary only in the σ factor associated with the core. In place of the normal σ factor (σ^{55}) they contain alternative σ subunits of various molecular weights. The *in vitro* transcription of isolated *spo* genes and certain vegetative genes, using the various forms of RNA polymerase, shows that each form initiates the transcription of a specific set of genes. Furthermore, addition of the various purified σ factors to RNA polymerase core enzyme shows that each directs RNA polymerase to transcribe a different class of promoters *in vitro* (Gilman and Chamberlin, 1983, and references therein; Johnson *et al.*, 1983). The promoter sequences to which RNA polymerase is directed by each of these σ factors are shown in Fig. 2. While the normal vegetative σ (σ^{55}) directs transcription from promoters resembling the *E. coli* consensus promoter, the other σ factors dramatically alter promoter sequence specificity. The *Bacillus* phage SPO1 also makes use of different σ factors to ensure the correct temporal expression of the early, middle and late genes. Thus one of the early genes encodes a σ factor (gp28) which directs transcription of the middle genes. Similarly, two of the middle genes encode factors gp33 and gp34 which direct expression of the late genes (Losick and Pero, 1981). The promoter sequences to which RNA polymerase is directed by each of these different σ factors are shown in Fig. 2. While the normal vegetative σ (σ^{55}) directs transcription from promoters resembling the *E. coli* consensus promoter, the other σ factors dramatically alter promoter sequence specificity.

Until very recently it was believed that alternative σ factors played no role in regulating gene expression in *E. coli*. However, an elegant study by Gross and her colleagues (Grossman *et al.*, 1984) has clearly demonstrated that expression of the heat shock genes (which are transcribed at elevated temperatures) requires an alternative σ factor, the product of the *htpR* gene. The sequence of the *htpR* protein shows considerable homology with the normal cellular σ factor, the *rpoD* gene product (Landick *et al.*, 1984). The *htpR* protein co-purifies with RNA polymerase when overproduced from a multicopy plasmid and in *in vitro* transcription systems will direct RNA polymerase core enzyme to initiate transcription at heat shock promoters but not at other promoters. Similarly, the normal cellular σ factor fails to direct transcription from heat shock promoters *in vitro* (Grossman *et al.*, 1984). The identification of an alternative *E. coli* σ factor opens up the possibility that others may also exist. The product of a T4 gene, gp55, can replace *E. coli* σ factor and specifically direct RNA polymerase to transcribe phage T4 late promoters (Kassavetis and Geiduschek, 1984).

σ	-35 region	-10 region
σ^{55}	T T G A C A	T A T A A T
σ^{gp28}	T . A G G A G A . . A	T T T . T T T
σ^{37}	G G . T . A A A	T A T T G T T T
σ^{28}	C T A A A	C C G A T A T

Figure 2 Promoters recognized by *Bacillus* σ factors: the nucleotide sequences of *Bacillus* promoters recognized by different σ factors are shown. σ^{55} is the major vegetative σ factor. σ^{37} and σ^{28} are also present in vegetative cells while σ^{gp28} is a product of the *Bacillus*-specific phage SPO1.

There have also been indications and speculations that the *ntrA* gene product, which is required for the expression of nitrogen-regulated promoters, may also be a σ factor; certainly, the consensus sequence for nitrogen-regulated promoters is very different from that for the normal *E. coli* consensus promoter (Dixon *et al.*, 1983). Thus it is possible that various sets of *E. coli* genes of related function are transcribed by RNA polymerase containing different σ factors.

3 Proteins which alter DNA–RNA polymerase interactions: positive and negative regulatory proteins

The most familiar form of gene regulation is that mediated by the binding of proteins to DNA. Positive regulatory proteins enhance the productive binding of RNA polymerase to promoters whilst negative regulatory proteins reduce this binding. Several proteins are autogenously regulated. For example, alanyl-tRNA synthetase can bind to its own promoter, inhibiting further transcription and thereby controlling its own synthesis (Putney and Schimmel, 1981). The binding of regulatory proteins to DNA at the appropriate operator sequences can be enhanced or inhibited by interactions with small molecules. Thus lactose prevents the binding of *lac* repressor to the *lac* operator sequence, while cAMP promotes the binding of the catabolite activator protein (CAP; also called CRP protein) to many different operators. Global expression of many different genes can thus be coordinately regulated by variations in the intracellular concentrations of small messenger molecules, or alarmones, such as cAMP. Other nucleotides, such as ZTP (5-amino-4-imidazole carboxamide riboside 5′-triphosphate) or AppppA, may also serve as alarmones in response to a folate deficiency or oxidative stress respectively (Bochner and Ames, 1982; Bochner *et al.*, 1984). However, the mechanisms by which these nucleotides might affect gene expression are unknown.

The identification and characterization of regulatory proteins, and their physiological role in the regulation of gene expression under varying environmental conditions, are well documented and the interested reader is referred to any of a number of excellent textbooks and reviews. As these studies primarily involved genetic and biochemical techniques, rather than recombinant DNA technology, they fall outside the scope of this article. The more recent and exciting results, which have been dependent on the power of genetic engineering, concern the details of the interactions between regulatory proteins and DNA and the mechanisms by which these interactions enhance or inhibit RNA polymerase function. Most, if not all, regulatory proteins bind to DNA as dimers, recognizing a short inverted repeat sequence in the DNA. The two most fruitful approaches to understanding the interactions between DNA, the repressor–activator proteins and RNA polymerase have been footprinting and the use of both *in vivo* and site-directed mutagenesis.

DNA footprinting is used to determine the binding sites on DNA of proteins by identifying those bases which are protected from enzymes (e.g. DNaseI) or chemical reagents (e.g. dimethyl sulphate) by the bound protein. Many laboratories have used these techniques to identify the DNA binding sites for several regulatory proteins as well as for RNA polymerase (see for example Siebenlist *et al.*, 1980, and Travers *et al.*, 1983). As an illustration of the rather interesting data which are now being derived from such methodology I will describe some recent studies on the interactions of the CAP with DNA and with RNA polymerase. These studies begin to shed some light on the mechanisms by which the cAMP–CAP complex stimulates transcription from catabolite-repressible promoters.

When glucose is absent from the environment, the levels of cAMP in the cell are elevated by activation of adenylate cyclase. cAMP is bound by CAP and the cAMP–CAP complex binds to DNA close to the transcription initiation points of various promoters, including those of the *lac* and *gal* operons, enhancing transcription (Ullman and Danchin, 1983). S1–nuclease mapping of the endpoints of *in vivo* mRNA shows that when cAMP–CAP binds to the *gal* operator it stimulates transcription from a specific start point P1. However, in the absence of cAMP–CAP transcription is initiated from an alternative promoter P2 located 5 bp upstream from P1 (Aiba *et al.*, 1981) (Fig. 3). cAMP–CAP not only

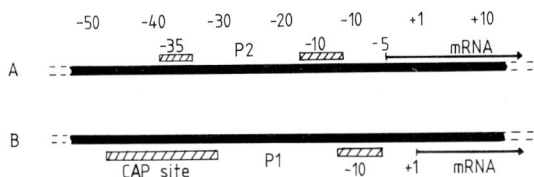

Figure 3 RNA polymerase and CAP binding at the *gal* promoter. The *gal* promoter region is numbered with the P1 transcription initiation point as $+1$. A, in the absence of the cAMP–CAP complex RNA polymerase interacts at the -10 and -35 regions of P2, as illustrated, initiating transcription at base -5. B, in the presence of cAMP–CAP the complex binds at the -35 to -45 position; RNA polymerase then binds to the DNA at P1, on the opposite side of the helix from P2, and initiates transcription at $+1$. Adapted from Spassky *et al.* (1984).

enhances transcription from P1 but also represses transcription from P2. Transcription from P2 is required even under catabolite-repressing conditions to ensure synthesis of the GalE protein required for cell wall biosynthesis; mRNA molecules initiated at P2 are translated differentially from those initiated at P1 such that only the GalE protein is produced (Queen and Rosenberg, 1981). Using DNase footprinting, Buc and his collaborators (Spassky *et al.*, 1984) have probed the sequences in the *gal* promoter which are protected by RNA polymerase, both in the presence and in the absence of the cAMP–CAP complex. These results show that RNA polymerase binds to a single site (site 2) in the absence of cAMP–CAP and initiates transcription from P2. In the presence of cAMP–CAP, RNA polymerase is prevented from binding to site 2 and binds to an alternative site (site 1), initiating transcription from P1. It is clear that the cAMP–CAP complex does not simply prevent the binding of RNA polymerase to site 2 but positively stimulates binding to site 1, as it stimulates the rate of open complex formation at P1. This is probably due to direct interactions between RNA polymerase and CAP. DNA footprinting shows that the spacing between the two proteins on the promoter–operator DNA is precisely that required for the two proteins to interact and, in addition, RNA polymerase can stimulate CAP binding to mutant *gal* promoters to which CAP will not bind on its own. Interestingly, once RNA polymerase is bound to site 2 it cannot be displaced by cAMP–CAP even though cAMP–CAP can still bind to the DNA. This again is compatible with the binding sites identified by footprinting: RNA polymerase bound at site 2 is on the opposite side of the DNA double helix from the cAMP–CAP binding site (Fig. 3).

The isolation of operator mutations has played a major role in the study of protein–DNA interactions and in defining those bases required for protein binding. Using again the example of CAP and the *gal* operator, Fig. 4 shows the bases which, if mutated, prevent CAP binding. These bases are believed to be those recognized by the CAP protein in the major groove of the DNA. Mutations which alter the amino acid sequence of the regulatory protein can define those amino acids involved in DNA recognition. Alteration of glutamic acid 181 in the CAP protein alters the specificity with which it binds DNA, changing its recognition site from AA-TGTGA--T---TCA-AT to AA-TGTAA--T-TCA-AT (Ebright *et al.*, 1984). This indicates that glu181 makes direct contact with base pairs 7 and 16 of the DNA binding site (each of the two subunits of the dimeric protein interacting with one of these bases).

Clearly, the most complete understanding of protein–DNA interactions will come from elucidation of the three-dimensional structures of

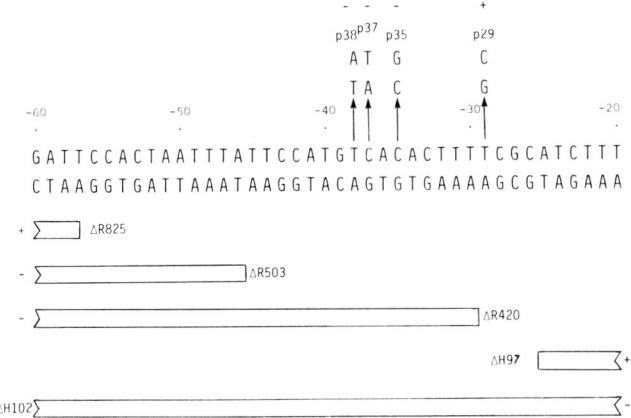

Figure 4 Mutations in the *gal* promoter which alter cAMP–CAP binding. The sequence of the *gal* promoter is numbered back from the point of transcription initiation (− 1). The extent of deletions within this region is shown by the bars underneath the sequence. Base changes introduced by point mutations within the promoter are shown above the sequence: −, mutations which prevent stable CAP binding; +, mutations which have no effect. From Kolb *et al.* (1983).

the complexes. The structures of three regulatory proteins have now been determined, those for the two λ repressor proteins cI and cro as well as that for the positive regulatory protein CAP (Anderson *et al.*, 1981; McKay and Steitz, 1981; Pabo and Lewis, 1982). These proteins all share structural similarities which probably reflect similar interactions between α helices in the proteins and the major groove of the DNA (Sauer *et al.*, 1982; Steitz *et al.*, 1982). In addition, one regulatory protein, the phage 434 repressor, has been cocrystallized with a synthetic operator oligonucleotide (Anderson *et al.*, 1984). Interestingly, DNA structure does not appear to be greatly perturbed by the binding of the repressor. This approach clearly holds great potential for aiding our understanding of protein–DNA interactions. Apart from facilitating the production of large amounts of material for structural studies, recombinant DNA technology has played a relatively minor role in such work and a detailed discussion is therefore outside the scope of this chapter. However, there is now considerable potential for a combined genetic and structural analysis of protein–DNA interactions using site-directed mutagenesis to alter specific interactions in the DNA–protein complex and to define the principles of recognition specificity. This is an area where major advances will almost certainly be made in the not too distant future.

B   Elongation of mRNA

During transcription the elongation of mRNA does not proceed at a constant rate but in a discontinuous fashion, RNA polymerase pausing at specific sites on the template. For example, during the *in vitro* transcription of phage T7 DNA, RNA polymerase is paused for about 70% of the time (Kassavetis and Chamberlin, 1981). Pause sites are heterogeneous, both in their efficiency and in their sequence. However, many pause sites coincide with inverted repeat sequences (see for example Winkler and Yanofsky, 1981). In addition, the stem–loop structures of ρ-independent terminators (section IIC) are thought to induce RNA polymerase pausing as a necessary first step in termination.

A role for pausing in the regulation of gene expression is now beginning to emerge. For example, the regulatory nucleotide ppGpp markedly enhances RNA polymerase pausing during transcription of both T7 and *rrnB* DNA (Kingston and Chamberlin, 1981; Kingston *et al.*, 1981). ppGpp is synthesized by *E. coli* in response to amino acid starvation and is required for the stringent response—the rapid cessation of synthesis of rRNA, tRNA and ribosomal proteins. It can be calculated that a pause of about 1 s is sufficient to reduce the maximum expression of an rRNA operon. *In vitro*, ppGpp can induce pausing at specific sites in the *rrnB* operon for periods of up to 20 s. If this also occurs *in vivo* it would result in a sixfold reduction in overall transcription rate. In addition, the *rrnB* operon has two promoters, P1 and P2, which initiate transcription 119 bases apart. At high RNA polymerase concentrations, RNA polymerase initiating from P1 *in vitro* is induced to pause for up to 4 min at a site 90 bp downstream. This pausing is enhanced by ppGpp. Not only does the pausing preclude further transcription from P1 but the paused polymerase also blocks initiation at the second promoter P2 (Kingston and Chamberlin, 1981). Because pausing is dependent on RNA polymerase availability this may provide a means of responding to varying RNA polymerase concentrations within the cell.

Unfortunately, little is known about the role of pausing *in vivo*. It has been suggested that it plays a role in the synchronization of transcription and translation necessary for efficient function of the *trp* attenuator sequence (Winkler and Yanofsky, 1981; see section II.C). Recent evidence, using synthetic oligonucleotides to inhibit stem–loop formation in the *trp* leader mRNA, has shown that inhibition of pausing by the disruption of secondary structures at the pause site relieves attenuation

(Fisher and Yanofsky, 1984). Thus at least in this case it seems probable that transcription pausing plays a role *in vivo*. It seems likely that this process will be found to play a part in determining the rate of transcription of many genes. However, until investigated in more detail, further roles for pausing, its role in coupling transcription to translation and its link to termination remain speculative.

C   Termination of transcription

The termination of transcription is a complex event which is by no means fully understood. Indeed, the more that termination is investigated, the more complex the phenomenon appears to be. Transcription termination signals can be broadly divided into two classes, the factor-independent and the factor-dependent terminators (see Holmes *et al.*, 1983, for a recent review).

1   *Factor-independent terminators*

Factor-independent terminators do not require the function of any proteins other than RNA polymerase; a direct interaction between the enzyme and a signal on the DNA template is sufficient to cause termination. Many factor-independent terminators have been characterized both *in vivo* and *in vitro*. They all have two features in common: a GC-rich inverted repeat sequence which is immediately followed by a number of Ts (Us in the mRNA) (Fig. 5). Termination occurs within this run of Ts. The inverted repeat is thought to form a stem–loop structure within the RNA which causes RNA polymerase to pause whilst the relative instability of the rU–dA base pairs between the template and transcript facilitates release. However, despite the apparent simplicity

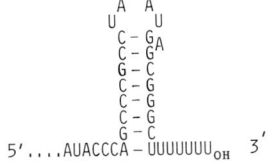

Figure 5 A typical factor-independent terminator is a GC-rich stem–loop followed by several U residues. This example is the terminator from the *trp* attenuator leader mRNA.

of such terminators, many questions remain to be answered. How does the stem–loop modify RNA polymerase such that it terminates transcription? How does the stem–loop structure exert its effect on an RNA polymerase molecule which, by necessity, must already have transcribed and passed over this sequence? What are the structural features of the terminator which determine its relative efficiency? How do proteins, particularly antitermination factors (see below), alter termination at these sites?

2   *Factor-dependent terminators*

Factor-dependent terminators constitute an extremely heterogeneous group of sites at which transcription termination occurs. Termination at these sites does not occur in *in vitro* systems unless specific protein factors are added. The most familiar of these protein factors is ρ. Although a consensus sequence for ρ-dependent terminators (CAATCAA) was originally derived from the characterization of three such sites, it is now clear that other ρ-dependent sites differ markedly from this consensus. In addition, ρ-dependent terminators often display considerable heterogeneity at the 3' end of the terminated RNA. It now seems likely that ρ-dependent termination does not depend on a specific nucleotide sequence so much as on a lack of structural features in the RNA, such as stable stem–loops. For ρ to function, single-stranded RNA must be exposed, free of bound ribosomes or other proteins. ρ is thought to bind to the exposed RNA and travel along the elongating chain in a 5'→3' direction until it catches up and interacts with the paused RNA polymerase molecule. Interactions between the two proteins then cause the polymerase to terminate transcription. This provides an explanation for the ρ-dependent polarity of gene expression caused by nonsense (chain termination) mutations. The nonsense mutation causes ribosomes to terminate translation prematurely, exposing naked RNA to which ρ can then bind and facilitate premature termination.

3   *Attenuation*

Termination of transcription is emerging as an important mechanism by which gene expression can be regulated. The clearest example of a regulatory role for termination is attenuation. In general terms, attenuation can be considered as the premature termination of transcription within an operon, the control of which modulates the expression of downstream genes. Attenuation appears in many guises, the most familiar of which is in the regulation of amino acid biosynthetic operons, so elegantly elucidated by Yanofsky and his coworkers. Indeed, the term attenuation is sometimes used in a restrictive fashion to refer only to this type of regulation. However, other mechanisms are known by which the efficiency of termination can be altered, in particular those involving antitermination proteins. Because these antitermination factors function in a very specific manner to influence termination, they will be considered separately below.

The attenuation mechanism, by which several amino acid biosynthetic operons are regulated, is familiar enough to be included in many basic textbooks. The model was first proposed by Yanofsky and his colleagues in the mid-1970s, being based primarily on the unusual sequence of the 5′ end of the *trp* operon mRNA. This model was so elegant and appealing that it rapidly became accepted. However, it took several years and many ingenious experiments before the many facets of the model could be tested and confirmed.

The basic mechanism of attenuation is outlined in Fig. 6, using the *trp* operon as an example. The *trp* mRNA has a long 5′ leader sequence between the start of the mRNA and the initiation codon for the first structural gene (*trpE*). This leader RNA contains several sequences of overlapping dyad symmetry which are capable of forming alternative secondary structures. The structure most distant from the promoter is a GC-rich stem–loop followed by eight uridine residues which, in the presence of excess tryptophan, functions as a terminator. However, when tryptophan is limiting, transcription proceeds through this site and into the structural genes of the operon. The proportion of RNA polymerase molecules which terminate at the attenuator stem–loop is determined by the extent to which ribosomes can translate a small 14 amino acid polypeptide encoded by the leader mRNA. As RNA polymerase transcribes the leader RNA, ribosomes bind and begin to translate the leader polypeptide. This leader polypeptide contains two tandem trp codons (in the leader peptide of the *his* operon there are seven tandem his codons) and, when tryptophan (or more accurately tryptophanyl–tRNA [trp]) is limiting, the ribosome stalls at these codons. The stalling of the ribosomes allows formation of the "pre-emptor" stem–loop in the mRNA. Once formed, the pre-emptor is stable and precludes formation of the terminator stem–loop with which it overlaps. Transcription thus proceeds into the *trp* structural genes. However, in the presence of excess tryptophan, ribosomes do not stall but proceed to the stop codon of the leader polypeptide. The ribosomes now overlap the dyad symmetry and therefore preclude formation of the pre-emptor

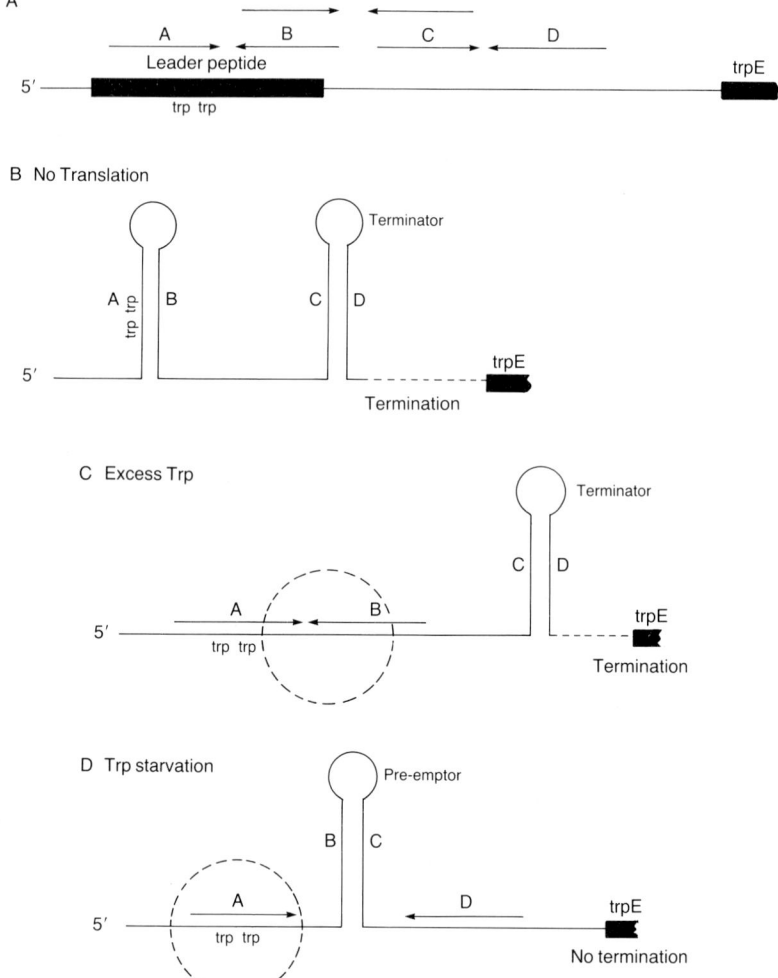

Figure 6 The general mechanism of attenuation in the *trp* operon is shown diagrammatically. A, The relative location of the coding region for the *trp* leader peptide (black bar) and the inverted repeat sequences (horizontal arrows) at the 5′ end of the *trp* operon are shown. B, In the absence of translation, stem–loop structures AB and CD can form and transcription is terminated. C, Excess tryptophan allows ribosomes (broken circle) to translate the entire leader peptide, preventing the formation of stem–loop AB. The terminator stem–loop CD can still form and transcription terminates. D, Tryptophan starvation causes ribosomes to stall at the trp codons in the leader peptide. Stem–loop AB is thus unable to form, allowing formation of the pre-emptor (stem–loop BC). This, in turn, precludes formation of the terminator (stem–loop CD) and allows transcription to proceed through the structural genes of the operon.

stem–loop. The terminator stem–loop is able to form, transcription is terminated and the structural genes of the operon will be neither transcribed nor translated.

The experimental evidence which supports this model comes from many different approaches, both biochemical and genetic. Several lines of biochemical evidence have been obtained over the last eight or ten years as molecular genetic techniques have advanced.

(i) *In vitro* transcription of *trp* leader DNA shows that the attenuator stem–loop functions as a strong terminator (Bertrand *et al.*, 1976). Similarly, mapping of the *in vivo* mRNA shows that in the presence of tryptophan most mRNA terminates at the attenuator stem–loop whilst under conditions of tryptophan starvation very little mRNA terminates at this site (Bertrand *et al.*, 1976).

(ii) Evidence that the leader peptide gene (*trpL*) is translated has been obtained by fusion of the *lacI* and *trpE* genes to the leader peptide in such a way that their products are synthesized using the translational initiation signals of *trpL*. The hybrid proteins produced by such fusions have amino termini identical with that predicted for the *trpL* leader peptide (Miozzari and Yanofsky, 1978).

(iii) Evidence that the predicted secondary structures can form, at least *in vitro*, has come from the use of RNase T1 to digest *trp* leader RNA synthesized *in vitro* (Lee and Yanofsky, 1977; Oxender *et al.*, 1979). Regions of the leader mRNA resistant to this enzyme correspond well to those sequences predicted to be involved in stem–loop formation.

(iv) Direct evidence for the importance of RNA secondary structure has been obtained recently using synthetic oligonucleotides complementary to various regions of the leader RNA. In *in vitro* transcription systems, oligonucleotides complementary to the pre-emptor prevent its formation and, as predicted by the model, enhance termination at the attenuator. Similarly, oligonucleotides complementary to the attenuator stem–loop prevent its formation and transcription no longer terminates (Fisher and Yanofsky, 1984). Use of similar oligonucleotides has also shown that the pausing of RNA polymerase is important for termination (see section II.B).

(v) RNA–RNA interactions, rather than RNA–DNA interactions, have been shown to be important in termination at the attenuator by elegant experiments in which mutations which destabilize the attenuator stem–loop have been introduced. The single-stranded M13 phage was used to construct heteroduplexes such that while one of the DNA strands carried the mutation the other strand contained the wild-type sequence. These mutations were found to affect termination at the attenuator only when present on the transcribed strand, so that the mutation is introduced into the leader mRNA (Ryan and Chamberlin, 1983).

A considerable amount of genetic evidence has provided strong support for many aspects of the attenuation model, for both the *trp* and the *his* attenuators. The most detailed genetic analysis has been carried out on the *his* operon (Johnston and Roth, 1981). Using various selections Johnston and Roth isolated and sequenced many classes of mutation which provide strong support for different aspects of attenuation.

(i) Nonsense mutations in the leader peptide, which can be suppressed by tRNA suppressor mutations, show that the leader peptide is translated into protein *in vivo*.

(ii) Chain termination mutations which result in the synthesis of a shortened leader peptide show that the extent to which this peptide is translated can control the proportion of mRNA molecules which terminate at the attenuator.

(iii) Mutations in the initiation codon or the ribosome binding site for the leader peptide which reduce its translation reduce operon expression, again showing that translation of the leader peptide is important.

(iv) Frameshift mutations, which allow the leader peptide to be produced but with a very different amino acid sequence, show that it is the rate of translation of this region of RNA, rather than the precise amino acid sequence of the leader peptide itself, which is important. Thus the leader polypeptide *per se* has no function.

(v) Mutations which reduce the stability of the terminator stem–loop increase operon expression; if the entire loop is deleted attenuation is absent. Similarly, all mutations which alter the stability of the alternate stem–loops have the predicted effects. For example, destabilization of the pre-emptor decreases operon expression. As formation of the pre-emptor stem–loop is thereby reduced, the attenuator forms preferentially and premature termination results.

Despite the elegant nature of the attenuation model and the formidable body of supporting experimental evidence, this mechanism of regulation is, by necessity, restricted to a small number of operons involved in the synthesis and utilization of amino acids. However, attenuation in its more general sense, the premature termination of transcription, could apply to many classes of operon. Several other examples of control by attenuation are beginning to emerge. For example, the chromosomal β-lactamase gene of *E. coli* is regulated by growth rate. At low growth rates transcription initiates at the promoter but terminates 40 bases downstream (Fig. 7). However, at high growth

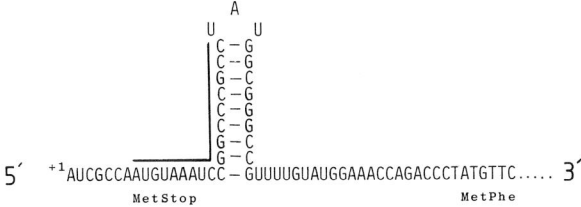

Figure 7 Attenuation of transcription in the β-lactamase gene. The sequence of the 5′ end of the *E. coli* β-lactamase mRNA is shown up to (and including) the first two codons of the β-lactamase structural gene. The attenuator–terminator stem–loop is shown. Ribosomes which initiate and terminate at the adjacent AUG and UAA codons (nucleotides 8–13) are believed to cover about 12 adjacent bases (——) which overlap the terminator loop and prevent it forming. At high growth rates, when ribosomes are abundant, termination does not occur. Only at low growth rates, when free ribosomes are not available, can the attenuator stem–loop form, causing termination of transcription and preventing expression of the operon.

rates termination does not occur and the gene is transcribed and translated efficiently. The formation of the attenuator stem–loop in this case seems to be controlled by the abundance of ribosomes. At high growth rates ribosomes bind to a site which overlaps the attenuator sequence, preventing stem–loop formation and consequently preventing termination. At low growth rates, however, the number of ribosomes per cell is reduced so that none binds to the ribosome binding site and attenuation occurs (Jaurin *et al.*, 1981). Evidence for terminators at the beginning of rRNA operons (Kingston and Chamberlin, 1981) and also within the operon which encodes two ribosomal proteins and the β subunits of RNA polymerase (Barry *et al.*, 1980) suggests that attenuation, in one form or another, operates in other operons too.

4 *Antitermination*

Antitermination was first discovered in phage λ (Adhya *et al.*, 1974). Transcription which initiates at the major promoters P_L or P_R terminates at the ρ-dependent sites t_{L1} and t_{R1} respectively (Fig. 8). The *N* gene product is encoded between P_L and t_{L1}. Once synthesized, the N protein functions as an antiterminator to prevent termination at both t_{L1} and t_{R1} and transcription of the rest of the λ genome can proceed. The λ *Q* gene product is also an antiterminator, acting at $t_{R'}$. In order for the N protein to function as an antiterminator, a specific 17 bp nucleotide sequence,

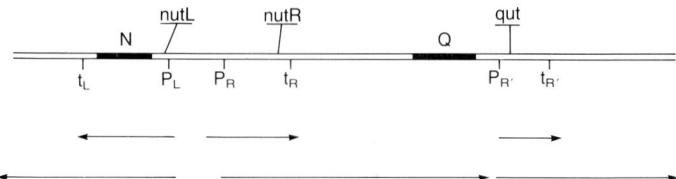

Figure 8 The location of the signals involved in phage λ termination–antitermination are shown diagrammatically. Terminated mRNA species are shown as the upper set of arrows. Antiterminated mRNA species, the result of N or Q action, are shown as the lower set of arrows.

nut, must be present between the promoter and the terminator (Rosenberg *et al.*, 1978). The *nut* sequence need bear no fixed positional relationship to the promoter or the terminator. It is believed that the N protein in some way modifies RNA polymerase as it transcribes the *nut* site, which prevents RNA polymerase from recognizing ρ-dependent termination signals. A similar sequence, *qut*, is required for Q-dependent antitermination. The *nut* and *qut* sites are specific to the N and Q proteins respectively. Other phages also encode specific antitermination proteins, each of which recognizes a sequence analogous to, but different from, *nut* and *qut*.

In addition to the N protein, certain chromosomally encoded proteins are required for efficient antitermination. One of these proteins, NusA, is well characterized. It is a 68 000 dalton protein which interacts directly with the N protein and also with RNA polymerase. In the latter case it competes for binding with the σ subunit (Greenblatt and Li, 1981a, 1981b). NusA is required for antitermination by all antiterminator proteins (N, Q etc.). A 10 bp sequence CGCTCT(T)TTAA, called box A, is thought to play a role in recognition by NusA (Olson *et al.*, 1982). Box A is conserved within the *nut*, *qut* and other equivalent sites. The variable part of the *nut* and *qut* sites presumably confers the specificity for the individual antiterminator proteins N and Q. The NusB protein is a 14 000 dalton polypeptide which, like NusA, is not terminator specific and is essential for the function of both N and Q (Ghosh and Das, 1984). Its mechanism of action is unknown. Mutations in the S10 ribosomal protein (NusE) can also influence antitermination, suggesting that interactions between the ribosome and RNA polymerase may also be important (Friedman *et al.*, 1981). The mechanisms by which NusA, NusB, ribosomes and the N or Q proteins interact with RNA polymerase to inhibit termination remain obscure.

In addition to their role in antitermination in phage λ, which led to their discovery, the NusA and NusB proteins also affect the expression of many chromosomally encoded genes. Thus, during *in vitro* transcription of the *trp* operon, termination of transcription at several sites within the operon is prevented in the presence of the NusA and NusB proteins (Kuroki *et al.*, 1982). NusA is involved in termination at *trpt* (Farnham *et al.*, 1982) and in the *rrnB* leader region (Kingston and Chamberlin, 1981) and can prevent termination in the middle of the *lac* and *rpoB* operons (Kung *et al.*, 1975; Zarucki-Schulz *et al.*, 1979). However, although NusA plays a wide variety of roles in altering transcription termination its physiological role in the regulation of chromosomal gene expression remains to be elucidated.

A possible regulatory role for termination at the end of operons is less clear. However, the fact that in the system which has been most intensively studied, i.e. the terminator of the *trp* operon (Holmes *et al.*, 1983; see also section III.A), tandem terminators have been identified, each influenced by different factors, suggests that termination may be subject to physiological control. The *tyrT* gene also has multiple terminators. Whether many genes possess multiple termination signals, whether these can interact and whether any possible regulatory roles can be ascribed to them remain open questions.

III   Regulation by RNA processing and degradation

The amount of any particular mRNA species in the cell will depend not only on the rate of synthesis but also on the rate at which it is degraded. It has been known for many years that the half-life of bacterial mRNA is very variable; some species of RNA, such as those encoding ribosomal components or outer membrane proteins, are relatively stable, whereas others have half-lives of 1 or 2 min. However, relatively little is known about the specific processes which determine mRNA degradation and thus might affect gene expression. This is due, at least in part, to the experimental difficulties inherent in studying bacterial mRNA: the very high activity of ribonucleases within the cell and the relatively short half-lives of most mRNA species *in vivo*. While a number of degradative ribonuclease activities from *E. coli* have been characterized in some detail (Kaplan and Apirion, 1974; Gegenheimer and Apirion, 1981), little is known about their specific roles *in vivo* or the factors which influence their activity (Apirion, 1973). Three main factors can influence mRNA stability: the secondary structure of the molecule itself, interactions between the mRNA and ribosomes and cleavage of mRNA by specific endonucleases.

A   mRNA secondary structure

Little is known about those features of mRNA secondary structure which determine its rate of degradation. However, it is a general observation that stable mRNA species can adopt a large number of stem–loop structures, particularly in their 3′ and 5′ untranslated regions. Evidence that stem–loop structures can serve as a block to exonuclease activity has been obtained for a number of operons. One interesting example of this is the terminator at the end of the *trp* operon (Fig. 9). A classical GC-rich stem–loop structure followed by a run of Us (*trpt*) (see section II.C) is found 3′ to the most distal gene of the operon. Mapping of the *in vivo* mRNA shows that the 3′ end of *trp* mRNA falls at the base of this stem–loop structure. However, this sequence was found to function inefficiently as a terminator *in vitro*. In addition, selection for mutations defective in termination at the end of the *trp* operon showed that most of the RNA polymerase molecules do not terminate transcription at *trpt* but at a ρ-dependent terminator (called *trpt′*) about 250 bp downstream (Wu *et al.*, 1980, 1981). Thus most RNA polymerase

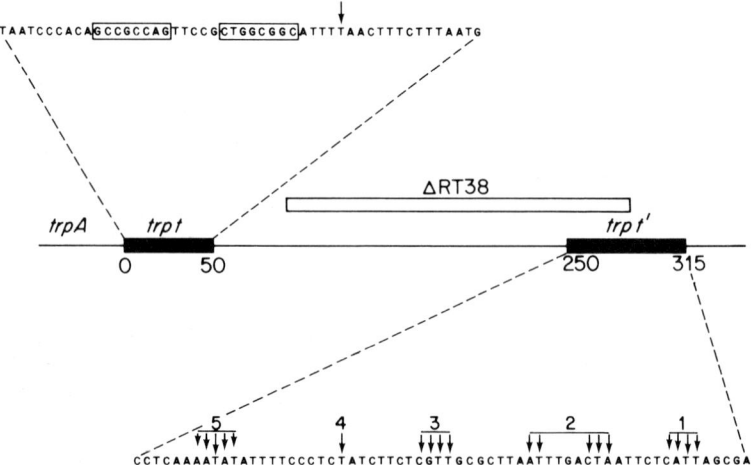

Figure 9 Tandem terminators in the *trp* operon. Immediately following the *trpA* gene is the factor-independent terminator stem–loop (*trpt*). *In vivo*, the majority of RNA polymerase molecules proceed through *trpt* and terminate about 200 bp downstream at *trpt′*. The arrows indicate the many sites at which σ-dependent termination occurs within *trpt′*. ΔRT38 is a deletion which allows readthrough of transcription at the end of the *trp* operon. From Wu *et al.* (1981).

molecules fail to terminate at *trpt* and proceed through to *trpt'*. mRNA terminated at *trpt'* is presumably attacked by $3' \rightarrow 5'$ exonucleases which processively digest the mRNA until they reach the GC-rich stem–loop at *trpt* where further degradation is blocked (Holmes *et al.*, 1983). This explains the observation that all *in vivo* mRNA has its endpoint at *trpt*. The function of the stem–loop of *trpt*, at least in part, is to protect the mRNA from degradation by processive exonucleases. Similarly, stem–loop structures formed by bacterial REP sequences, present in a wide variety of operons, appear to act as a partial block to $3' \rightarrow 5'$ exonucleases (Stern *et al.*, 1984a). The untranslated regions of most mRNA species contain potential stem–loop structures. It seems likely that the role of many of these is protective, serving to control the rate at which that particular mRNA molecule is degraded by processive exonucleases.

The only really detailed study of mRNA degradation at a molecular level has resulted from an elegant procedure for determining the half-life of different regions of any given transcript (von Gabain *et al.*, 1983). This procedure, which should be generally applicable, involves the digestion of a uniformly labelled single-stranded DNA probe, complementary to the entire transcript, into discrete fragments with one of the several restriction endonucleases which cleave single-stranded DNA. The resulting fragments are then hybridized to total cellular RNA and treated with S1–nuclease and the protected fragments are separated by gel electrophoresis. If initiation of RNA synthesis is inhibited by rifampicin, mRNA can then be isolated from cells after varying time intervals and the pattern of decay examined. Using this procedure it is possible to obtain accurate half-lives for any mRNA species and, in addition, to determine whether all parts of the mRNA are degraded at the same rate or at different rates. Evidence has been obtained in this way which indicates that *ompA* mRNA, which is a relatively stable species, is degraded in a $3' \rightarrow 5'$ direction, the 5' end of the mRNA having a longer half-life than the 3' end (von Gabain *et al.*, 1983). Interestingly, the half-life of certain mRNA species was found to exhibit a marked dependence on growth rate. Thus it seems that the rate of mRNA degradation can be specifically modulated as a means of adjusting gene expression, by some mechanism as yet unknown (Nilsson *et al.*, 1984). This is a field where much remains to be investigated.

B   Effects of protein binding on mRNA degradation

The binding of protein to mRNA molecules clearly could protect them from degradation by ribonucleases. However, with the exception of ribosomes, little is known about the possible role of such processes *in vivo*. There is a considerable amount of evidence, accumulated over many years, that the rate of mRNA degradation can be affected by translation. Intuitively one might anticipate that the absence of ribosomes would result in increased mRNA degradation and, indeed, there are several reports to this effect (e.g. Schneider *et al.*, 1978, and Yates and Nomura, 1981). However, there are also apparently conflicting reports that the presence of ribosomes can decrease the half-life of mRNA (e.g. Har-El *et al.*, 1979). It has been suggested that nuclease activity is associated with the ribosomes themselves and thus, under certain circumstances, ribosomes enhance degradation. Despite the evidence that, at least in some cases, the presence of ribosomes on an mRNA molecule protects it from nuclease attack, ribosomes cannot protect untranslated regions of mRNA. Any possible role for proteins other than ribosomal proteins also remains obscure. With the plethora of recombinant DNA techniques now available this is an area ripe for investigation.

C   Endonuclease cleavage of mRNA

In eukaryotic cells one of the more controversial mechanisms by which gene expression might be regulated is the removal of introns from precursor RNAs. The absence of introns from eubacteria precludes regulation by this means. Or does it? Chu *et al.* (1984) have demonstrated the presence of an intron in the thymidylate synthase gene of phage T4 and it has recently been shown that this intron is removed from the precursor RNA by a post-transcriptional processing event (Belfort *et al.*, 1985). The indications that splicing can occur for one mRNA molecule (whether it be enzyme mediated or self-processing) imply that it might also occur for others. Clearly most, if not all, bacterial genes lack introns. If splicing does occur it is therefore likely to be restricted to a limited set of genes. In addition, self-processing of mRNA as a regulatory mechanism is at least a possibility. It has now been convincingly demonstrated that certain RNA species have the ability to cleave themselves at a specific point (Kruger *et al.*, 1982). A single example of self-cleavage has recently been demonstrated in prokaryotes, the cleavage of T4 species 1 RNA (Watson *et al.*, 1984). However, the biological role of this event is not yet clear and neither is the generality of such processes in bacterial cells. The possible regulatory significance of these intriguing findings remains to be elucidated.

Although the breaking and rejoining of mRNA is limited in bacteria, cleavage of mRNA by specific endonucleases may serve an important regulatory function. Many endoribonucleases which are involved in processing rRNA and tRNA have been characterized in *E. coli* (Gegenheimer and Apirion, 1981). Some of these nucleases may also cleave mRNA. There is a considerable body of rather non-specific, but convincing, data that mRNA species undergo functional inactivation prior to degradation (Shen *et al.*, 1981). This might well be the result of endonuclease activity, cleaving the mRNA and exposing it to exonucleases. Many multicistronic operons are apparently cleaved to monocistronic mRNA species as a first step in degradation (Achord and Kennell, 1974; Blundell and Kennell, 1974; Schlessinger *et al.*, 1977) and functional inactivation of *lac* mRNA is apparently associated with RNaseIII (Shen *et al.*, 1981). However, at a molecular level little is known about these processes. S1–nuclease mapping of mRNA endpoints has indicated that endonucleolytic cleavage occurs in a number of mRNA species, although the enzymes involved are generally unknown. However, cleavage of mRNA by RNaseIII can play an important role in modulating gene expression (Gitelman and Apirion, 1980; Robertson, 1982). The two best examples of regulation by mRNA cleavage are the cleavage of phage T7 RNA and of λ *int* mRNA by RNaseIII. In these two cases the mechanisms by which gene expression is altered by the cleavage event are very different.

The λ *int* gene can be transcribed from either of two promoters, the major leftward promoter P_L or a minor *int*-specific promoter P_I (Fig. 10). Transcription initiating at P_I terminates at a site t_{int} just downstream from the *int* gene while transcripts initiating at P_L proceed through this terminator as a result of N-dependent antitermination at *nutL* (see section II.C). Mutations at a site *sib*, 240 bp 3′ to the *int* gene and overlapping t_{int}, increase expression of *int* from P_L but not from P_I (when present in *cis*). Such a regulatory role for 3′ sequences is termed retroregulation. Sequence analysis of a large number of *sib* mutants shows that they all fall within a region of dyad symmetry and destabilize the formation of a potential stem–loop structure (Guarneros *et al.*, 1982). As the structure of *sib* resembles known RNaseIII processing sites, and as *sib* has no effect on *int* expression in RNaseIII-deficient strains, it seemed likely that processing of *int* mRNA by RNaseIII plays a role in retroregulation. Purified RNaseIII has been shown to cleave the *sib* site of mRNA synthesized *in vitro*. The mechanism by which cleavage of *int* mRNA can alter the expression of the upstream gene appears to be by an enhancement of the rate of degradation of upstream mRNA (Rosenberg and Schmeissner, 1982). Oligonucleotide fingerprint analysis of the P_L

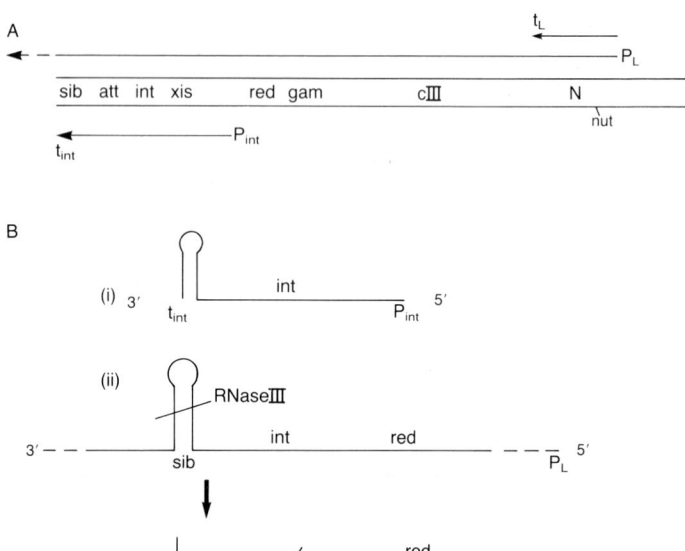

Figure 10 Regulation of λ *int* expression. A, Map of part of the left arm of phage
λ showing the arrangement of elements involved in *int* expression. B, Probable
mechanism of *int* expression. (i) mRNA initiating at P_{int} terminates at t_{int}. The
stem–loop of the terminator protects the mRNA from degradation by $3'{\rightarrow}5'$
exonucleases. (ii) RNA polymerase initiating at P_L is modified at the antiterminator site (*nut*) so that it does not terminate at t_{int} but proceeds to transcribe the
remainder of the left arm of λ. Transcription through t_{int} extends the length of
the stem–loop structure so that it is now a substrate for RNaseIII (the *sib* site).
Cleavage of mRNA at this site exposes the upstream *int* mRNA to degradation
by $3'{\rightarrow}5'$ exonucleases, thus specifically reducing *int* expression from the P_L
promoter.

transcript shows that, while both upstream and downstream oligonucleotides are present in equimolar amounts, oligonucleotides characteristic of the *int* gene itself are present in dramatically reduced amounts.
In accordance with this model, the *int*-specific oligonucleotides are
found to be present in an RNaseIII-deficient mutant. Thus it appears
that RNaseIII cleavage renders upstream mRNA, including the *int*
gene, susceptible to exonucleolytic degradation. mRNA initiating at P_I
is not degraded, presumably because the stem–loop provided by t_I
(absent from the P_L transcript) blocks $3'{\rightarrow}5'$ exonuclease activity.

The second, and very different, example of regulation by RNaseIII
processing is provided by phage T7. It has been known for several years

that the early mRNA of T7 is cleaved by RNaseIII at five separate sites. In at least one case this cleavage event affects the expression of genes encoded by the mRNA. Genes 1.1 and 1.2 are expressed from the same mRNA molecule produced by cleavage of T7 mRNA at two sites R4 and R5 (Fig. 11). Certain mutations which alter the expression of genes 1.1 and 1.2 map downstream of the two genes in RNaseIII site R5 (another example of retroregulation). The R5 site is actually cleaved by RNaseIII at two points *in vivo* designated sites A and B (Fig. 11). While site B is cleaved to completion, site A is normally cleaved in only about 40% of the transcripts, giving rise to two mRNA species with slightly differing 3' ends. Mutations which alter the stability of the stem–loop structure of site R5 alter the proportion of molecules which are cleaved at site A and thus the relative proportions of the two mRNA species (Saito and Richardson, 1981). The smaller transcript is able to express proteins 1.1 and 1.2 while the larger transcript cannot. This can be explained by

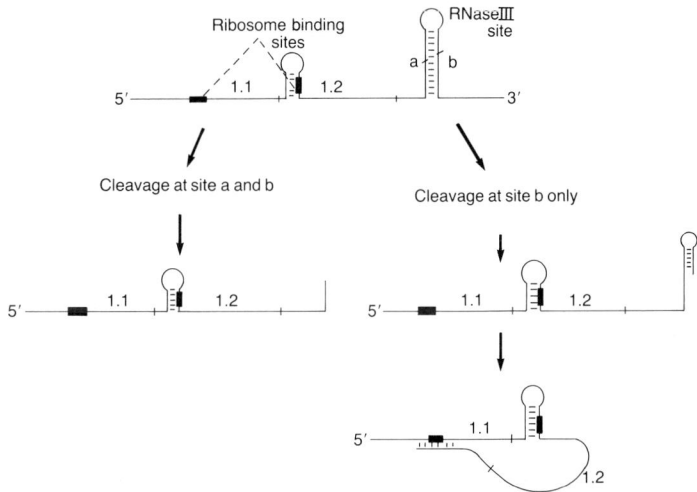

Figure 11 Regulation of T7 gene expression by RNaseIII cleavage. RNA encoding proteins 1.1 and 1.2 can be processed in one of two alternative ways by RNaseIII. Processing at both sites a and b results in a short mRNA species: gene 1.1 is translated; the passage of ribosomes along gene 1.1 prevents formation of the stem–loop structure which normally occludes the ribosome-binding site of 1.2 and thus allows 1.2 to be translated as well. Alternatively, the mRNA is cleaved only at site b: this results in a longer mRNA species, complementary to the ribosome-binding site of gene 1.1 which inhibits translation of gene 1.1. As translation of gene 1.2 depends on prior translation of gene 1.1, this longer mRNA species also fails to express gene 1.2. Adapted from Saito and Richardson (1981).

homology between the unique 3′ end of the larger mRNA species and the ribosome-binding site of gene 1.1 (Fig. 11). The longer mRNA species can fold back on itself and inhibit translation; this cannot occur with the shorter transcript and thus gene 1.1 is expressed. Gene 1.2 is also only expressed from the shorter mRNA species as its translation is dependent on the prior translation of gene 1.1; the ribosome-binding site of 1.2 is occluded by a stem–loop structure which is disrupted by ribosomes translating gene 1.1.

I have described two very different examples of endonuclease cleavage events which alter gene expression; in one case the cleavage event promotes selective degradation of mRNA. RNaseIII is known to cleave other bacterial mRNA species (e.g. *lac* and *rplJ*), although the function of these cleavage events is unclear. It seems very likely that RNA processing and the control of mRNA degradation play a major role in determining the level of bacterial gene expression. With the many available techniques, and the many tantalizing observations described above, this is an area where answers should soon be forthcoming.

IV   Regulation by DNA rearrangement

The bacterial genome is not static but can undergo a variety of structural rearrangements. While many of these rearrangements are relatively random events which only alter gene expression in an uncontrolled fashion, others have evolved as precise and specific mechanisms for regulating gene expression.

A   Random rearrangements

Most DNA rearrangements are essentially random events and, although the sites at which the rearrangements take place may show some sequence specificity, they affect gene expression only fortuitously. Such events include duplication, insertion and transposition.

Duplications are frequent events in bacteria and up to 3% of the chromosome is duplicated at any one time (Anderson and Roth, 1981). The duplications are unlikely to serve any specific regulatory role although they are often selectively maintained when pressure is exerted for overexpression of a particular gene. However, certain genes are maintained in a permanently duplicated state on the chromosome. These tend to be the genes which must be expressed at high levels. For example, there are two distinct copies of the gene encoding translation elongation factor Tu and in the case of the rRNA genes seven non-tandem copies are maintained. Presumably the increased gene dosage assists in the high expression of these operons.

Many insertion sequences (IS) are present in *E. coli* and it is well documented that the insertion of an IS into a gene will inactivate that gene (see Calos and Miller, 1980, and Kleckner, 1981, for recent reviews). Similarly, IS elements often contain promoters directing transcription outwards and into the flanking chromosomal DNA. Consequently, when inserted upstream of a gene an IS can provide an alternative promoter for its transcription and thereby alter expression or regulation (see for example Jaurin and Normark, 1983). A good example of this is the *bgl* (*β*-glucoside utilization) operon of *E. coli* which is not expressed in many isolates. Mutations which allow expression of the operon arise spontaneously and at high frequency. These mutations are usually insertions of either IS1 or IS5 upstream of the operon which provide promoter function (Reynolds *et al.*, 1981). However, it is not clear whether these IS insertions provide a new promoter or activate a nearby chromosomal promoter by altering the local DNA structure and/or supercoiling (see section II.A). It has been suggested that the random switching of this operon is of selective advantage. While many *β*-glucosides serve as useful carbon sources, others are toxic. If a population of cells can randomly switch the expression of the *bgl* operon on and off, a proportion of the cells will always survive should toxic *β*-glucosides be present in the environment.

Transposition of genes around the chromosome is apparently a rare event; the gene order on the chromosomes of *E. coli* and *Salmonella typhimurium* is highly conserved. This is very probably because gene position on the chromosome is of functional significance (Roth and Schmid, 1981) and might be for one or both of the following reasons.

(i) The chromosome is probably separated into 40 or so domains, each of which is independently supercoiled (Sinden and Pettijohn, 1981). As the level of supercoiling is very carefully regulated, and can affect gene expression (see section II.A), chromosomal position may well alter expression.

(ii) Because the chromosome replicates bidirectionally from a single origin, regions of the chromosome in the vicinity of the origin will be present in more copies per cell than genes located near the terminator of replication. The position of a gene on the chromosome will therefore determine its relative copy number and hence expression.

Although transposition of genes to different sites on the chromosome does not seem to be an important regulatory mechanism in *E. coli*, transposition of genes to expression sites is well known for the expression of alternative surface antigens in eukaryotic microorganisms such as *Trypanosoma* and may well also play a role in bacterial species which exhibit similar antigenic variation.

B   Controlled rearrangements

The best example of DNA rearrangement as a specific mechanism for controlling gene expression is the case of phase variation in *S. typhimurium*. Any single cell of *S. typhimurium* expresses one of two alternative flagellar proteins H1 or H2, which are encoded by genes located in different points on the chromosome. Cells switch from expressing one gene to the other in an apparently random fashion and at a relatively high frequency in order to avoid the host immune response. The switching event is dependent on sequences located adjacent to the H2 gene (Fig. 12). Restriction analysis of phage carrying this region of the chromosome shows that the DNA immediately upstream of H2, isolated either from strains expressing H1 or from strains expressing H2, can exist in two alternative forms, differing from each other by the inversion of a specific segment of DNA (Silverman *et al.*, 1979; Silverman and Simon, 1980). This DNA fragment is 995 bp long and is bounded by a 14 bp inverted repeat sequence which is essential for the inversion process (Zieg and Simon, 1980). The 995 bp fragment includes a promoter which transcribes outwards from the invertible region and into the adjacent chromosomal DNA. In one orientation this results in the expression of H2 and an adjacent gene which encodes a repressor of H1 synthesis (Fig. 12). When in the inverted orientation neither H2 nor the H1 repressor can be synthesized and hence the H1 antigen is produced. Inversion of the 995 bp fragment is mediated by a polypeptide encoded by the *hin* gene which is within the invertible region.

Other genetic switches also involve related inversion events. For example, the host range of bacteriophage Mu is controlled by the invertible G region, first identified as a loopout in electron micrographs of DNA heteroduplexes between variants with different host ranges. Unlike phase variation, the invertible segment of Mu does not contain the gene encoding the transposase (*gin*) or the promoter but encodes the structural genes for the tail fibres which determine host range (Giphart-Gassler *et al.*, 1982). Inversion of this segment places different tail fibre genes under the control of a promoter which directs transcription into the invertible region (Fig. 12). DNA inversion also controls phage P1 host range (controlled by the *cin* gene product; Iida *et al.*, 1983). An *E.*

A   Orientation 1 (H2 expressed)

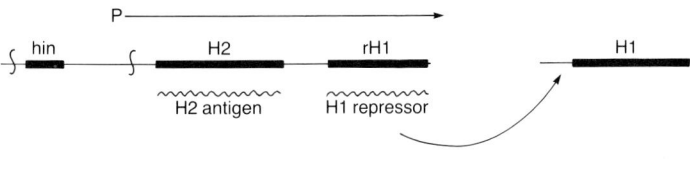

Orientation 2 (H1 expressed)

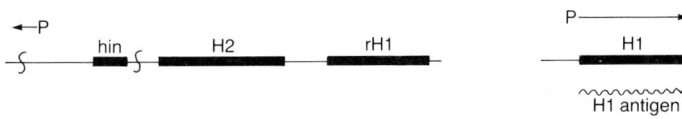

B   Orientation 1 (S and U expressed)

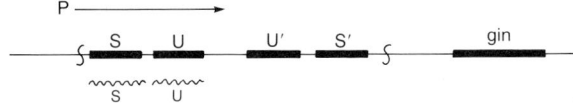

Orientation 2 (S′ and U′ expressed)

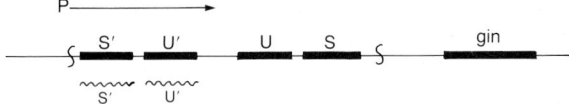

Figure 12 DNA inversion. The arrangements of elements involved in *Salmonella* phase variation (A) and Mu host-range determination (B) are illustrated diagramatically. Thick black bars represent protein coding regions. The invertible DNA segments are delineated by ∫. P indicates promoters and adjacent arrows show the direction and extent of transcription. See the text for a detailed explanation. (Not drawn to scale.)

coli gene, *pin*, has also been identified which mediates the inversion of an 1800 bp chromosomal DNA fragment, the function of which is unknown (Plasterk *et al.*, 1983). Interestingly, the *hin*, *gin*, *cin* and *pin* gene products all share sequence homology with each other and with the *tnpR* gene product of transposon Tn3 (Plasterk *et al.*, 1983)). The recombination sites recognized by these various site-specific recombination systems also share sequence homology. Thus it seems very probable that they operate by similar mechanisms and evolved from a common origin. This raises the "chicken and egg" question: did the chromosome "steal" and adapt the site-specific recombination systems from transposons for its own use, or did the recombination systems evolve on the chromosome and later "escape" as transposable elements?

In addition to the above examples, there is some evidence that the expression of alternative restriction and/or modification systems may also be mediated by gene rearrangements and that this might involve a cassette-type model analogous to that of yeast mating-type control (Glover *et al.*, 1983). Complex mechanisms such as the mammalian immunoglobulin rearrangements, *Trypanosome* antigenic variation or yeast mating-type cassettes, have not yet been characterized in *E. coli*. Nevertheless, the possibility that variations on these types of re-arrangement occur in *E. coli*, or in other bacterial species, is quite likely and should by no means be dismissed.

V   Translational regulation

The preceding sections describe the various mechanisms by which *E. coli* regulates the cellular levels of individual mRNA species. However, it should be obvious that the total amount of each protein synthesized by the cell will depend not only on the amount of mRNA which encodes it but also on the efficiency with which that particular mRNA species is translated into protein. Translational efficiency will depend both on the efficiency with which ribosomes can bind to the mRNA and initiate translation and on the rate at which translation proceeds. Only recently have many of the subtleties associated with these two processes begun to emerge. It is becoming clear that translation can be modulated by many factors and is likely to play an important role in determining gene expression. These factors can be divided into two classes: (i) signals encoded within the mRNA molecule itself and (ii) regulatory molecules which interact with the mRNA and alter its translational efficiency.

A   Signals encoded within the mRNA

The most straightforward means by which translational efficiency can vary between mRNA species are those which are encoded by the mRNA molecule itself: the "strength" of the ribosome-binding site and the relative frequency with which different codons are used.

1   *Ribosome-binding sites*

The ribosome-binding site, or Shine–Dalgarno sequence, is the signal which directs the ribosomes to bind to the mRNA and to initiate translation at the correct AUG codon. It consists of a short purine-rich sequence, complementary to the 3′ end of 16S rRNA, which is located between 5 and 10 bp upstream from the initiation codon (Fig. 13) (Shine and Dalgarno, 1974). The 16S rRNA is believed to base-pair with this sequence prior to translational initiation. Variations in the efficiency of recognition and binding of ribosomes to these sites will, of course, lead to variations in gene expression.

Ribosome-binding sites were first identified by comparison of the nucleotide sequences immediately upstream from the initiation codons of a variety of genes. A detailed analysis of known ribosome-binding sites has been published (Stormo *et al.*, 1982). Despite the large number of bacterial genes which have now been sequenced, the criteria for defining a ribosome-binding site are still rather poorly defined. As few as three bases complementary to 16S rRNA are sufficient to confer ribosome-binding capacity. However, sequences other than the Shine–Dalgarno sequence, which have not yet been defined, must also play a role in directing ribosome binding. In addition, the codons GUG and UUG can sometimes be used as initiation codons instead of AUG; the significance of this is not known. It is also not understood why some sequences which are complementary to the 3′ end of 16S rRNA do not

```
5'...GGAUCACCUCCUUA        3'          16S  rRNA
            | | | | |   OH
            | | | | |
5'..........GGAGG...4-9bp...AUG..3'    mRNA
```

Figure 13   Shine–Dalgarno sequence. The sequence of the 3′ end of 16S rRNA, which is believed to interact directly with mRNA, is shown. The bases in mRNA complementary to 16S rRNA which are most frequently conserved are indicated. Generally, at least three adjacent bases from the sequence GGAGG are conserved.

function as sites for translational initiation. Consequently it is not yet possible to predict with any certainty the strength of a ribosome-binding site from its sequence alone. Indeed, it seems that some sites which show extensive homology with 16S rRNA are actually rather poor initiation sites; presumably the ribosome is so tightly bound that release and extension of the polypeptide chain is significantly reduced.

A comparison of the primary sequences of ribosome-binding sites does not take into account other factors which might affect translation initiation. For example, mRNA secondary structure in the vicinity of a ribosome-binding site can significantly increase or decrease its efficiency (see for example Cannistraro and Kennell, 1979, Queen and Rosenberg, 1981, Schwartz et al., 1981, Singer et al., 1981, and Hall et al., 1981). In the case of retroregulation in phage T7 (see section III), and in several ribosomal protein operons (see Fig. 14), ribosome-binding sites are inaccessible to ribosomes unless secondary structures are prevented from forming by ribosomes actively translating upstream genes.

Besides secondary structure, the position of a gene in a multicistronic operon can also be important. In many multicistronic operons genes expressed in a 1:1 ratio are separated by very short intergenic regions. Indeed, the initiation codon of one gene can overlap the termination codon of the preceding gene. The possibility that the genes in an operon can be translationally coupled (i.e. that ribosomes pass directly from one gene to another without leaving the mRNA) was first demonstrated by analysis of nonsense mutations in the *trpE* gene (Oppenheim and Yanofsky, 1980). These mutations were found to be more strongly polar on the expression of *trpD*, which is immediately downstream from *trpE*, than on expression of the more distal genes of the *trp* operon. It seems that mutations which cause translation of *trpE* to terminate prematurely also prevent ribosomes from initiating at *trpD*, presumably because the ribosomes are now unable to pass directly from the termination codon of *trpE* to the initiation codon of *trpD*. Translational coupling between the *galT* and *galK* genes has been convincingly demonstrated using *in vitro* techniques to construct a series of plasmids in which defined translation termination codons have been introduced at varying distances from the initiation codon of the *galK* gene (Schumperli et al., 1982). The efficiency with which *galK* is translated was found to depend to a considerable extent on the position at which translation of the upstream *galT* terminates; the more distant the termination codon of *galT* is from the initiation codon of *galK*, the less efficiently *galK* is translated.

Although translational coupling presumably occurs in most multicistronic operons, in those operons in which the genes are expressed at very different levels translational coupling must be relatively unimportant and expression of the various genes depends instead on the relative efficiencies of their individual ribosome-binding sites. A good example of this is the *unc* operon which encodes the many subunits of ATPase. The various subunits are translated in very different amounts from the same mRNA molecule, the position of the genes on the mRNA bearing no obvious relationship to their relative expression.

2   Codon usage

Once translation is initiated, the rate at which protein is synthesized may depend on codon usage. Certain codons for a given amino acid could be translated more efficiently than others. Codon usage is certainly non-random and appears to reflect the relative abundance of the various isoaccepting tRNA species (Ikemura, 1981). It is also a general observation that highly expressed genes include very few codons which are recognized by minor isoaccepting species of tRNA while poorly expressed genes tend to have a higher proportion of these rare codons (Konigsberg and Godson, 1983). In addition, codon context (i.e. the nature of the surrounding bases) can affect the efficiency with which that codon is translated (Bossi and Roth, 1980). However, despite such circumstantial evidence, there is as yet little or no direct evidence which demonstrates that codon usage plays a significant role in determining translational efficiency.

It is perhaps relevant to mention here that tRNA molecules contain many modified nucleosides and that the extent of this modification can dramatically affect the efficiency with which the tRNA species involved can function in translation (Eisenberg *et al.*, 1979; Turnbough *et al.*, 1979). The modification of several tRNA species can vary specifically under different environmental conditions. For example, hydroxylation of the isopentenyl group of 2-methylthio-N^6-(D2-isopentenyl)adenosine in the anticodon of many tRNA species does not occur under anaerobic conditions (Buck and Ames, 1984) and methylthiolation of various tRNAs is altered by the presence or absence of iron (Buck and Griffiths, 1982). There is evidence that the iron-dependent methylthiolation of tRNA can affect the expression of aromatic biosynthetic and transport operons (Buck and Griffiths, 1982) and that queuosine modification of tRNA can specifically alter the expression of nitrate reductase (Janel *et al.*, 1984). However, the intriguing possibility that differential tRNA modifications play a major role in regulating gene expression remains to be fully established.

B   Regulation by masking of ribosome-binding sites

Any molecule which can bind to a specific mRNA species in the vicinity of its ribosome-binding site could potentially regulate translation of that mRNA by inhibiting ribosome binding. Over the last four or five years several very different examples of such a mechanism have been uncovered. The best example is the elegant feedback mechanism by which ribosomal proteins bind specifically to their own mRNA and prevent further translation. In addition, the last year or so has brought forward two exciting examples of small regulatory RNA molecules which interact with the ribosome-binding sites of specific mRNA species and inhibit their translation.

1   *Regulation of ribosomal protein synthesis.*

The ribosome is a complex of many species of protein and RNA the syntheses of which are coordinately regulated. The various ribosomal components are maintained in the cell in the correct ratios so that excess proteins which cannot be assembled into mature ribosomes are not synthesized. These ratios are achieved by an autogenous feedback mechanism whereby one key protein from each ribosomal operon acts as a regulator, both of its own synthesis and of that of other genes within the operon. There is now considerable evidence that this regulation is achieved by modulation of the efficiency with which mRNA encoding the ribosomal proteins is translated (for a comprehensive review see Nomura *et al.*, 1984).

The first evidence for post-transcriptional regulation of ribosomal protein synthesis came from the observation that when mRNA encoding ribosomal proteins is overproduced from multicopy plasmids the corresponding proteins are not necessarily overproduced. A DNA-dependent *in vitro* transcription–translation system was used to show that regulation took place by autogenous feedback inhibition of translation. For example, the synthesis of plasmid-encoded ribosomal proteins L11 and L1 (encoded by the two genes of the L11–L1 operon; Fig. 14) is inhibited by the addition of purified L1 protein to the *in vitro* system (Yates *et al.*, 1980). In addition, overproduction of L1 *in vivo*, achieved by placing the gene under the control of another promoter, leads to the inhibition of L11 synthesis (Dean and Nomura, 1980) and mutations which prevent the synthesis of L1 stimulate the synthesis of L11 (Jinks-Robertson and Nomura, 1981). The site of action of L1 was

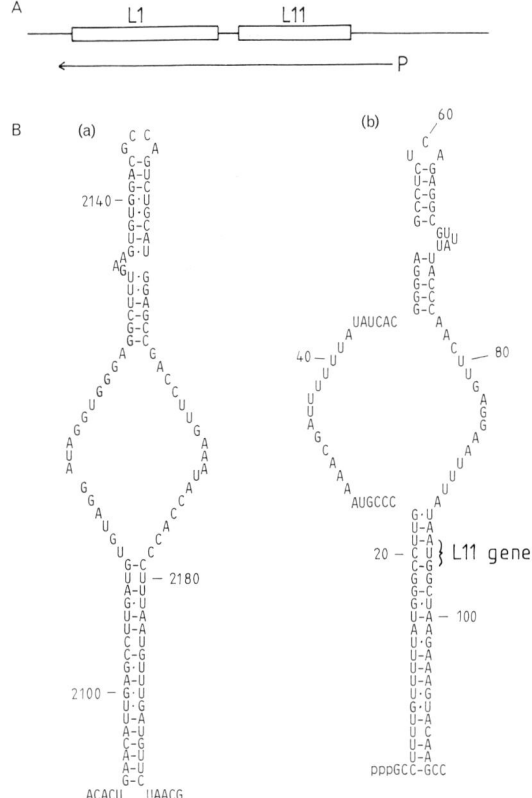

Figure 14   A, The organization of the L11–L1 operon. B, Structural similarities between (a) the L1-binding site on 23S rRNA and (b) the 5′ end of L11–L1 mRNA. Sequences are from Baughman and Nomura (1984).

shown to be within the first 160 bases of the L11–L1 mRNA by a study of the inhibitory effects of purified L1 protein on the *in vitro* translation of plasmids into which various deletions had been introduced (Yates and Nomura, 1981). The sequence of the L11–L1 mRNA immediately upstream (5′) from the L11 initiation codon, to which the L1 protein binds, can adopt a secondary structure very reminiscent of the structure of 23S rRNA to which L1 protein binds within the ribosomal complex (Fig. 14). *In vitro* mutagenesis using synthetic oligonucleotides has been used to confirm that it is indeed this structure which is important for recognition by the L1 protein (Baughman and Nomura, 1984). Several other ribosomal operons are regulated in a very similar manner, and in each case the binding site on the appropriate mRNA molecule which is recognized by the regulatory ribosomal protein also strongly resembles the corresponding binding site for the same protein on rRNA (Nomura et al., 1980) (Fig. 15). Thus, the mechanism of regulation seems to be as follows. Both rRNA and the regulatory sites on ribosomal protein mRNA adopt similar secondary structures which function as binding sites for the regulatory ribosomal proteins. However, the affinity of the proteins for rRNA is greater than that for the regulatory sites on the mRNA. When there is an excess of ribosomal protein with respect to rRNA, this protein cannot be incorporated into ribosomes and is then able to bind to the regulatory site on the corresponding mRNA species and inhibit further translation. Most autoregulatory ribosomal proteins bind to the mRNA in such a way that they directly occlude the ribosome-binding site. However, the L10–L7/L12 complex binds to a regulatory site on the rplJ mRNA some distance upstream from the Shine–Dalgarno sequence. In this case the regulatory protein complex acts only indirectly, facilitating the formation of alternative secondary structures around the ribosome-binding site which preclude further translation (Christensen et al., 1984). It is very hard to imagine a more effective yet simple means of maintaining the correct ratios of each of the ribosomal proteins within the cell.

2   Inhibition of translation by complementary RNA

During the past year two intriguing examples of translational inhibition by small regulatory RNA molecules have been discovered, regulating expression of the Tn10 transposase and the relative levels of the two major outer membrane porins.

The transposon Tn10 encodes a transposase, the synthesis of which must be tightly controlled in order to prevent runaway transposition. Simons and Kleckner (1983) have shown that synthesis of transposase from a chromosomally encoded Tn10 is inhibited in trans by a multicopy plasmid carrying IS10 (the insertion sequence which comprises the ends of Tn10 and which encodes all the functions required for transposition and its regulation). This phenomenon is called multicopy inhibition (MCI). All but 75 bp of IS10 can be deleted without removing its capacity to exert MCI. Thus, intact transposase protein is not required for inhibition of its own synthesis. The 75 bp of DNA required for MCI include a promoter (pOUT) which directs transcription in the opposite direction to transcription of the transposase gene (which initiates from a promoter pIN), producing an RNA molecule complementary to the 5′

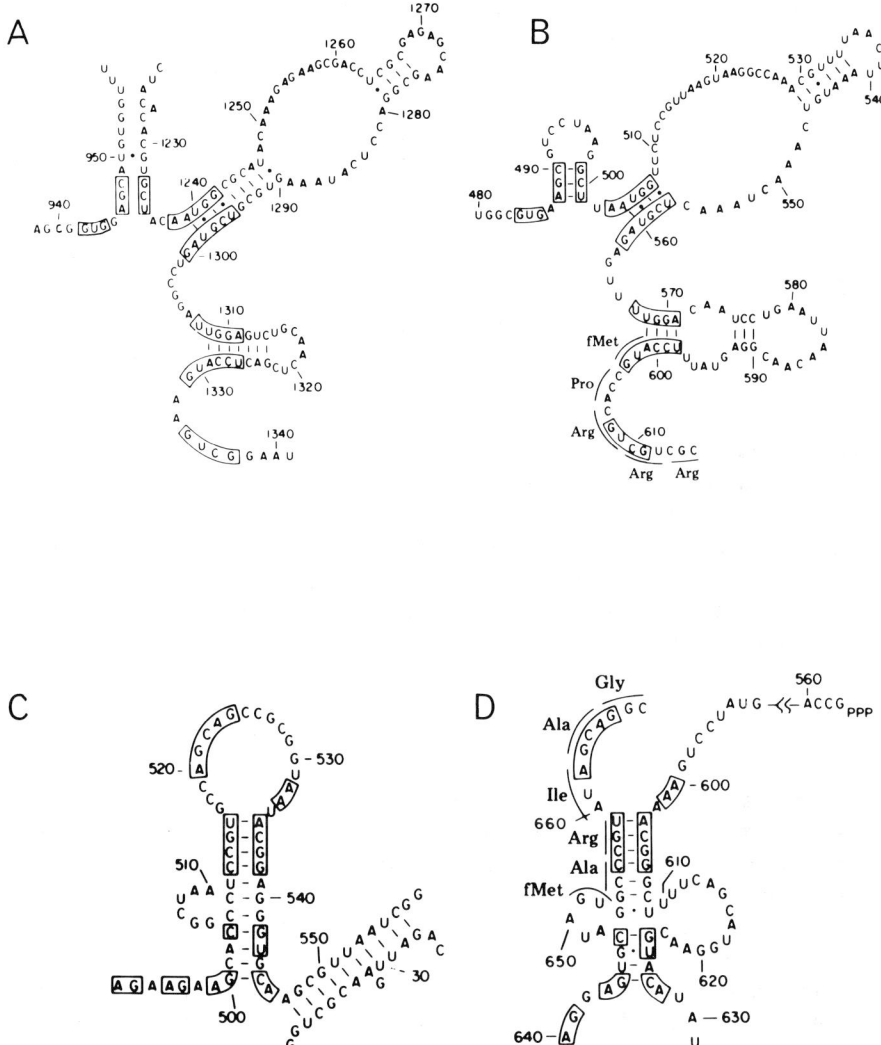

Figure 15 Similarities between ribosomal protein binding sites on mRNA and rRNA. A, B, Ribosomal protein S7 binding sites on 16S rRNA (A) and mRNA from the S12–S7 (*str*) operon (B). Boxed sequences indicate homology. The S12 coding region ends at base 501 and the S7 coding region begins at base 604. C, D, Ribosomal protein S4 binding sites on 16Sr RNA (C) and mRNA from the S13–S11–S4–α–L17 operon (D). Boxed sequences indicate homology. The S13 coding region begins at base 654. The figures are taken from Nomura *et al.* (1980).

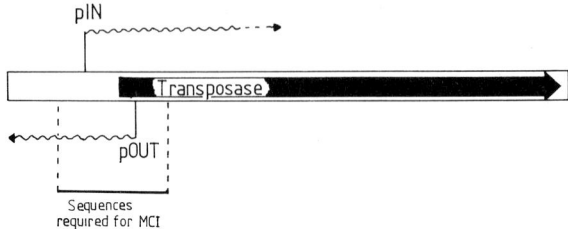

Figure 16 The structure of IS*10*-right, the insertion sequence which consti-
tutes one terminus of Tn*10*, is shown diagrammatically and not to scale. The
features involved in MCI are indicated.

end of transposase mRNA (Fig. 16). To determine whether MCI was due
to transcriptional or translational regulation, operon and gene fusions
(see section VI) were constructed *in vitro* between the transposase gene
and *lacZ*, such that transcription of *lacZ* was directed by the transposase promoter. In the case of gene fusions, in which translation of a
transposase–β-galactosidase hybrid polypeptide initiates at the normal
transposase initation codon, MCI of β-galactosidase activity was
observed. However, in operon fusions in which *lacZ* is translated from
its own ribosome-binding site, MCI of β-galactosidase synthesis was no
longer apparent. As *lacZ* was transcribed from the transposase promoter in both cases, while only in the gene fusions was β-galactosidase
translated from the transposase ribosome-binding site, the regulation of
transposase expression by MCI must be translational. Presumably, the
pOUT RNA inhibits translation of transposase by complementary base
pairing with the 5′ end of transposase mRNA, overlapping and therefore
occluding the translational initiation signals.

The second example of translational inhibition by antisense RNA is
found in the complex regulation of porin synthesis. The two porins,
encoded by the *ompC* and *ompF* genes which are located in different
places on the chromosome, are two of the most abundant proteins in *E.
coli*, responsible for the passage of nutrients through the outer
membrane. The relative expression of the two porins depends on the
osmotic composition of the external medium. However, the sum total of
the two proteins remains constant. During their characterization of the
ompC promoter, Mizuno *et al.* (1984) observed that a DNA fragment
upstream from the *ompC* promoter was able to inhibit *ompF* expression
when present on a multicopy plasmid. This fragment was shown to
encode a 175 base RNA species by Northern blotting and by S1–nuclease
mapping. The nucleotide sequence of this 175 base RNA molecule is

complementary to the 5′ end of *ompF* mRNA, including the ribosome-binding site. It therefore seems probable that this regulatory RNA molecule (named micRNA: mRNA-interfering complementary RNA) binds to *ompF* mRNA and inhibits the translation of the OmpF protein. By constructing *lacZ* fusions to the micRNA promoter it was shown that micRNA is expressed whenever *ompC* is expressed. Thus, when *ompC* is transcribed, micRNA is also produced, preventing OmpF protein synthesis. Similarly, when the *ompC* gene is not expressed, micRNA is not synthesized and *ompF* can be translated normally. A very subtle mechanism thus exists for maintaining a constant total amount of OmpF and OmpC proteins in the membrane.

It has not yet been directly demonstrated that RNA–RNA hybrids actually form in either of the cases described above and thereby inhibit translation. However, the construction of artificial micRNA molecules *in vitro*, complementary to the 5′ ends of a variety of different genes, shows that in all cases production of this antisense RNA specifically inhibits translation of the corresponding mRNA (Coleman *et al.*, 1984). Inhibition of expression of a eukaryotic gene has also been shown to be inhibited by an antisense RNA molecule encoded by a suitably constructed plasmid (Izant and Weintraub, 1984). Thus, although the general use of regulatory antisense RNA molecules *in vivo* has yet to be established, it is clear that such a mechanism provides considerable scope for the artificial manipulation of gene expression in both prokaryotes and eukaryotes.

VI   Gene and operon fusions

Most of the bacterial genes which have been studied in detail have been amenable to such studies because the gene product is relatively simple to assay. Much is known about the regulation of genes encoding enzymes such as those involved in sugar utilization or in amino acid biosynthesis. However, when the gene product is difficult (or impossible) to assay, study of the regulation of gene expression becomes much more difficult. This is true for many genes such as those encoding regulatory proteins, structural proteins with no enzymatic function (such as cell wall components) and proteins involved in complex functions such as transport across the membrane, as well as enzymes with inconvenient assays. The expression of such genes would be very much simpler to investigate if their promoter could be linked to a gene encoding an easily assayable enzyme in such a way that the promoter determines the activity of the enzyme. Several ingenious methods have now been developed for both the *in vivo* and the *in vitro* construction of such fusions. Most make use of the *lacZ* gene, encoding β-galactosidase. Although this is partly for historical reasons, *lacZ* is eminently suitable for constructing such fusions. Not only does the gene encode a stable enzyme for which there is a very simple assay but, in addition, the first 26 amino acids of β-galactosidase can be removed and replaced by almost any number of unrelated residues without inactivating the enzyme.

The terms gene fusion and operon fusion are often a source of confusion (Fig. 17). An operon fusion (or transcriptional fusion) is constructed so that *lacZ* is translated as an independent gene forming part of a polycistronic operon. The product of a *lacZ* operon fusion is native β-galactosidase, transcription of which depends on the promoter to which the gene is fused, but translation of which is directed by its own ribosome-binding site and initiation codon. Gene fusions (also called translational or protein fusions) differ in that a hybrid gene is constructed, the product of which is a hybrid polypeptide derived in part from the protein product of the gene under study and in part from β-galactosidase. In this case β-galactosidase activity (retained by the hybrid polypeptide) is translated from the initiation codon and ribosome-binding site of the gene being studied and not from those of the

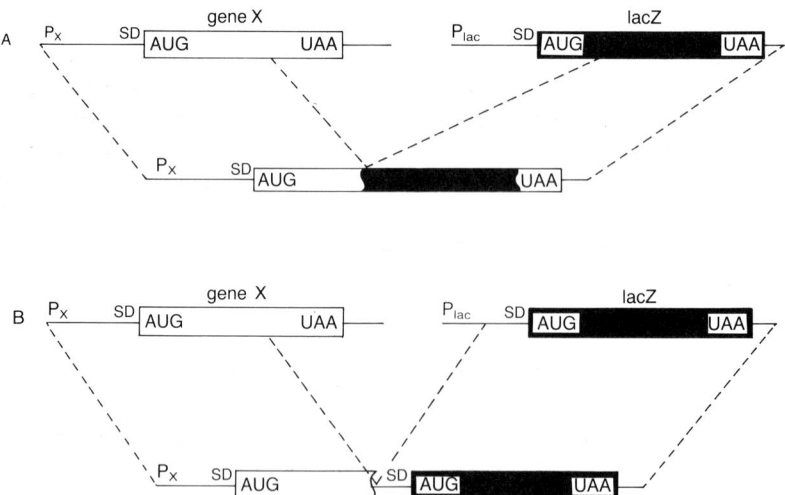

Figure 17 Gene (A), and operon (B) fusions. The distinction between a gene and an operon fusion is illustrated: P, promoters; SD, Shine–Dalgarno sequence; the thick bars represent coding sequences.

lacZ gene itself. To a first approximation, operon fusions provide an indication of transcriptional regulation while gene fusions provide a measure of regulation at both the transcriptional and the translational level. Gene fusions, which produce hybrid polypeptides, have an additional use in that they also provide a means of identifying and purifying otherwise intransigent proteins (see for example Shuman and Silhavy, 1981).

A   Construction of gene and operon fusions *in vivo*

In order to study the regulation of an operon in its normal chromosomal location (as opposed to in a non-physiological state on a plasmid) it is necessary to construct gene fusions *in vivo*. The first experiments making use of *β*-galactosidase fusions involved semirandom approaches, using transducing phage to place the *lacZ* gene on the chromosome near to the gene to be studied. Spontaneous deletions could then be selected which fused the two genes (see for example Miller *et al.*, 1970, Muller-Hill and Kania, 1974, Mitchell *et al.*, 1975, and Casadaban, 1976). Unfortunately, while providing much useful information, the construction and analysis of fusions obtained by these approaches required rather sophisticated genetic procedures which were not available to most laboratories. Until about five years ago the use of fusions was therefore rather limited. However, recently several refined and directed methods have been developed which enable gene and operon fusions to be constructed *in vitro* or *in vivo*, making it possible to fuse any gene to *lacZ*, even when that gene's function is unknown. These methods require little genetic expertise and can be used in any laboratory. Their general utility and usefulness is demonstrated by the increasing number of publications appearing each month using gene fusions to analyse the expression of every conceivable class of bacterial gene.

1   *Mu dI* (Amp, lac) *and its derivatives*

The simplest and most widely used method for constructing operon fusions was developed by Casadaban, making use of the properties of the transposable bacteriophage Mu (Casadaban and Cohen, 1979). On infection into a bacterial cell, bacteriophage Mu behaves as a transposable element, integrating its DNA into the bacterial chromosome essentially at random. Mu insertions into any gene of interest can be

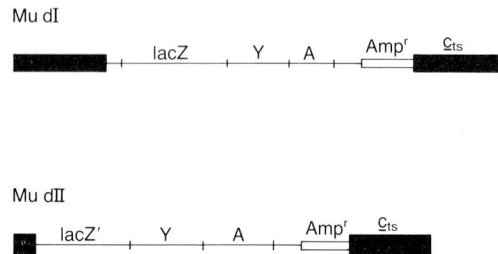

Figure 18   The structures of Mu dI and Mu dII are shown: ▬, Mu DNA. Mu dI contains the entire *lacZ*, *lacY* and *lacA* genes, lacking a promoter. Mu dII has all but 116 bp of one of the Mu termini deleted, as well as the first seven codons of the *lacZ* gene. In both phages the ampicillin resistance gene (Amp^r) is transcribed from its own promoter, independently of the external promoters which are required for transcription of the *lac* operon. Each phage has the c_{ts} repressor mutation.

readily isolated, assuming that a phenotype for identifying such mutations exists.

Phage Mu dI (Casadaban and Cohen, 1979) is a derivative of phage Mu which has been modified in several specific ways (Fig. 18). Firstly, parts of the genome of Mu which are not essential for its transposition functions have been removed. Secondly, a temperature-sensitive repressor mutation (c_{ts}) has been introduced so that the phage DNA becomes stably incorporated into the host bacterial genome at 30 °C but is induced to enter the lytic cycle when the temperature is raised to 37 °C. Thirdly, the *lac* operon, lacking its *in vivo* promoter, has been inserted into the phage. The *lac* DNA is positioned such that, when the Mu dI phage inserts into a gene in the correct orientation, transcription from that gene's promoter will proceed through the terminal Mu DNA sequences and into the *lacZ* and *lacY* genes conferring a Lac$^+$ phenotype. Finally, the β-lactamase gene from Tn*3*, conferring ampicillin resistance, has been incorporated into the phage genome.

It is a relatively simple matter to obtain Mu dI insertions into the chromosome. A strain lysogenic for the phage is induced at 37 °C to prepare a phage lysate. This lysate can be used to infect *E. coli* cells at 30 °C, selecting for ampicillin-resistant colonies. Each of the Amp^r colonies must be a Mu dI lysogen, the Mu phage having inserted at random in the chromosome. A large pool of many thousand independent insertions can be made in one day. Given that the *E. coli* genome contains about 4000 genes, a pool of at least 10 000 independent insertions should be obtained in order to be reasonably sure of obtaining at least one insertion in each gene. The Amp^r colonies are pooled and plated on appropriate selective plates in order to identify insertions within the gene of interest. Because the Mu dI phage can insert into this gene in one of two possible orientations, only one of which will place the *lacZ* gene under promoter control, only about 50% of the insertions isolated in any one gene will be Lac$^+$. However, all Lac$^+$ insertions will now have β-galactosidase expression under the control of the promoter of the gene being studied.

The ease with which Mu dI insertions can be isolated has resulted in widespread use of this phage to construct operon fusions. However, two major problems can be encountered when using this phage. Firstly, cells containing a Mu dI insertion are temperature sensitive owing to inactivation of the c_{ts} repressor at 37 °C which results in induction of the Mu replication and *kil* genes. In addition, the phage is still capable of transposition at a significant frequency (10^{-3}–10^{-4}). When used to select regulatory mutations which alter the expression of β-galactosidase from the fusion, the frequency of transposition becomes significant and a high proportion of mutants are a result of transposition to an additional site on the chromosome. Secondly, when a Mu dI fusion is transduced into a virgin (i.e. Mu-free) strain, zygotic induction results in a high frequency of transposition of the Mu to other sites on the chromosome. This makes mapping of Mu fusions difficult. In addition, the fusion cannot be easily introduced into different genetic backgrounds to facilitate the study of the effect of regulatory mutations on expression of the fusion. Putative regulatory mutations must always be introduced into the fusion strain, often a difficult proposition if there is no simple selection for transfer.

These problems can to some extent be overcome by stabilizing Mu dI insertions, selecting for derivatives of a fusion in which the transposition functions have been deleted. This has generally been achieved by selection for temperature-resistant derivatives at 42 °C (see for example Barrett *et al.*, 1984, and Stern *et al.*, 1984b). Such derivatives often contain deletions of a sizeable portion of the Mu genome including the function required for transposition. However, most involve deletions or other rearrangements of both Mu and adjacent chromosomal DNA and must therefore be carefully characterized to ensure that they still retain the original *lacZ* operon fusion. This procedure must also be undertaken separately for each independent fusion, a time-consuming process.

To overcome the problem of instability, a derivative of Mu dI has been constructed which has a Tn*9* (Cm^r) insertion in the Mu *B* gene. This polar mutation prevents expression of both the *B* and *kil* genes and therefore eliminates temperature sensitivity and, as the *B* gene is required for Mu replication, reduces the frequency of secondary transposition (Baker *et al.*, 1983). The phage Mu dX either can be used in place of Mu dI or, by homologous recombination, can be used to convert existing Mu dI insertions to more stable Mu dX derivatives. However, while more stable than Mu dI, Mu dX still suffers the problem of zygotic induction when transduced into a Mu-repressor-free background. Zygotic induction can, to some extent, be overcome by the use of a plasmid expressing the wild-type Mu *c* (repressor) gene (Krueger and Walker, 1983). However, this requires transformation of all recipient cells with the plasmid, a procedure which is often impractical.

A derivative of Mu dI has recently been devised which overcomes all these disadvantages. This phage, Mu dI-8, is stable and is not subject to zygotic induction when introduced into a virgin background (Hughes and Roth, 1984). Mu dI-8 is a derivative of Mu dI which has had an amber mutation introduced into the transposase gene. Thus, in a strain harbouring an amber suppressor (*supD, supE* or *supF*) the phage is able to transpose. A pool of random Mu dI-8 insertions can therefore be made in a suppressor strain and insertions in the particular gene required selected from this pool. However, when introduced by transduction or conjugation into a wild-type genetic background (i.e. with no suppressor mutation) the phage is unable to make functional transposase and can only integrate into the recipient chromosome by recombination. Zygotic induction is eliminated and the fusion can be transduced from strain to strain at will. In addition, in a suppressor-free background the Mu dI-8 phage is far more stable than Mu dI, reverting to Lac$^+$ at a frequency of less than 10^{-8}. Mu dI-8 should now be used routinely in preference to Mu dI.

Mu dI and its derivatives have been engineered for the formation of operon fusions. However, if translational regulation is to be studied, or the production of hybrid proteins is required, gene fusions can be constructed by similar one-step procedures. Thus Mu dII301 is a derivative of Mu dI which can be used in an identical fashion for the one-step construction of gene fusions (Casadaban and Chou, 1984). This phage is constructed so that when inserted into the chromosome only 117 bp of Mu DNA are located between the point of insertion and a truncated *lacZ* gene which lacks its first seven codons (Fig. 18). If inserted in the correct orientation and reading frame, transcription and translation from the gene into which the phage is inserted will proceed through the Mu DNA and into *lacZ*, producing a hybrid polypeptide. Because β-galactosidase is active even if the first seven amino acids are replaced, the hybrid polypeptide will normally possess β-galactosidase activity. A derivative of Mu dII301 with an amber mutation in the transposase gene (similar to that in Mu dI-8) has also been constructed (Hughes and Roth, 1984).

2   Use of λplac*Mu*

An alternative strategy for constructing gene fusions has recently been developed using a λ phage derivative (Bremer *et al.*, 1984). This phage, λplac*Mu*I, contains those sequences from bacteriophage Mu which enable it to integrate randomly into the chromosome by the Mu transposition mechanism. In addition, the phage contains the *lacZ* gene which has had its promoter, ribosome-binding site and first seven codons deleted. The truncated *lacZ* gene, like that of Mu dII301 (see above), is positioned so that, when inserted into the chromosome in the correct reading frame, transcription and translation from the gene into which it has been inserted proceed through the Mu DNA sequences and into the *lacZ* gene, resulting in synthesis of a hybrid polypeptide.

λplac*Mu*I contains a functional Mu *A* gene and is therefore able to transpose when infected into a recipient strain. However, transposition is stimulated when the Mu *B* gene is provided in *trans* by a helper phage (λpMu507). As a number of λ genes are deleted from λplac*Mu*I, including the *att* site, the phage cannot form lysogens by λ-dependent mechanisms. Thus a pool of random Lac[+] insertions is normally generated by double infection of the recipient with both λplac*Mu*I and the λpMu507 helper. Approximately one in six of all random insertions will be in the correct orientation and reading frame so that a *lacZ* hybrid protein is produced. A detailed experimental manual describing the use of λplac*Mu*I has recently been published (Silhavy *et al.*, 1984). λplac*Mu*I insertions are stable and can therefore be used to select regulatory mutations. In addition, they can be transduced from strain to strain without significant transposition by zygotic induction. λplac*Mu*I therefore has all the advantages of Mu dI-8. In addition, one further property of λplac*Mu*I can be particularly useful. Because the phage contains most λ functions, specialized transducing phages derived by imprecise excision can be selected after ultraviolet induction. Some of these specialized phages will carry the gene fusion and therefore provide a simple means of cloning the promoter and 5' regulatory sequences of the gene in question. Derivatives of λplac*Mu*I which can form operon fusions rather than gene fusions have also been constructed (E. Bremer, personal communication). A λplac*Mu*I which contains a kan[r] marker, to facilitate a selection and scoring of insertions, is also available.

3   Use of Mu dI and λplacMuI derivatives in species other than E. coli

While the construction of gene and operon fusions can, in theory, be applied to any species, the use of the various vectors for one-step construction of such fusions is limited by the host ranges of the phage on which the vectors are based.

Phage Mu and its derivatives will not infect *S. typhimurium*, *Klebsiella* or even *E. coli* B, let alone more distantly related species. However, various procedures have been devised for moving Mu derivatives into species other than *E. coli*.

(i) Strains can be selected which render them sensitive to Mu infection. While this has been successful in the case of *S. typhimurium* (Faelen *et al.*, 1981) it is unlikely that much success will be achieved with more distantly related species.

(ii) Helper phages which alter the Mu host range can be used. A Mu derivative with P1 host range can be used as a helper to assist Mu dI infection of *S. typhimurium* strains carrying a *galE* mutation which makes them P1 sensitive (Csonka *et al.*, 1981).

(iii) A more generally useful approach is to introduce the Mu phage into the host species by a procedure other than direct infection. This could be by transduction with a transducing phage with an extended host range or by transformation–conjugation with a broad host-range plasmid containing the Mu fusion vector. Once introduced into a species, the phage can be moved from strain to strain and induced to transpose to new locations by zygotic induction, by any transduction or conjugation procedures available for that particular species.

While little use has been made of *in vivo* gene fusions in species other than *E. coli* and *S. typhimurium*, there seems to be no reason why similar technologies should not be developed in other bacterial species.

B   Construction of gene and operon fusions *in vitro*

The use of *in vivo* fusions, present in a single copy on the chromosome, provides the best way to study the physiological regulation of any given gene and to identify, characterize and clone the regulatory genes which control its expression. However, the study of regulatory events at a molecular level requires the isolation of the genes and particularly their regulatory sequences. Because many gene products are difficult to assay, and because many genes are lethal when present on a multicopy plasmid, it is often advantageous to clone regulatory sequences in such a way that they control the expression of an easily assayable gene. Furthermore, various regulatory elements such as promoters, operators, terminators and ribosome-binding sites can be readily separated and studied independently. A whole host of cloning vectors have been constructed for the characterization of regulatory signals, often for very specific purposes. I shall here describe the basis of some of the more generally useful vectors to illustrate the types of information which can be obtained and any possible pitfalls.

The five basic features that are required for a useful promoter cloning vector are as follows.

(i) An origin of replication is necessary. Low copy number plasmids are more useful when studying the physiological regulation of plasmid-encoded genes *in vivo*. However, it is better, if possible, to use chromosomal, rather than plasmid, fusions for such studies. High copy number plasmids, based on the ColE1 origin of replication, are most frequently used, both for historical reasons and because of the ease of manipulation. However, for studies of expression in species other than *E. coli*, origins must be used which can replicate in the appropriate species.

(ii) A positive selection for retention of the plasmid is essential. This is normally a gene encoding a drug resistance enzyme such as β-lactamase (Amp^r) or chloramphenicol acetyl transferase (Cm^r).

(iii) A means of determining plasmid copy number is essential. As will be discussed below, plasmid copy number can vary considerably, depending on the size of inserted DNA, the strength of any inserted promoter and the genetic background, and must always be measured. This can be done on a comparative basis by "dot–blot" hybridization or more conveniently by assaying a plasmid-encoded gene product expressed independently of any inserted promoter fragment. As β-lactamase is not easy to measure, Cm^r is often the simplest marker to use.

(iv) The main feature of these vectors is an easily assayable gene that lacks its own promoter so that it is only expressed when an additional promoter is cloned in front of it. The two most frequently used genes are *lacZ* (β-galactosidase) and *galK* (galactokinase). These genes have easily assayable products and, in addition, there are several simple selections for mutations which increase or decrease their expression, facilitating the isolation and identification of regulatory mutations (Miller, 1972; McKenney *et al.*, 1981).

(v) Suitable cloning sites must be present in front of the *lacZ* or *galK* genes into which a promoter DNA fragment can be inserted. While many of the early vectors contained only single restriction sites, many have now had a synthetic multilinker oligonucleotide inserted to simplify the cloning of fragments obtained using a wide variety of restriction enzymes.

1   Vectors based on β-galactosidase

Vectors for the *in vitro* fusion of promoters to β-galactosidase were originally developed by Casadaban and Cohen (1980). These vectors, based on the multicopy ColE1 origin of replication, carry a selectable ampicillin resistance marker and the β-galactosidase gene lacking its promoter. Generally the *lacY* gene, encoding the lactose permease, is also present 3′ to *lacZ*. Restriction fragments can be cloned into sites located upstream from *lacZ* and those which carry a promoter in the correct orientation will direct β-galactosidase synthesis (Fig. 19). Many variants of the original Casadaban and Cohen vectors have been constructed in other laboratories.

These vectors can be used to determine whether a restriction fragment possesses promoter activity or for isolating promoter-contain-

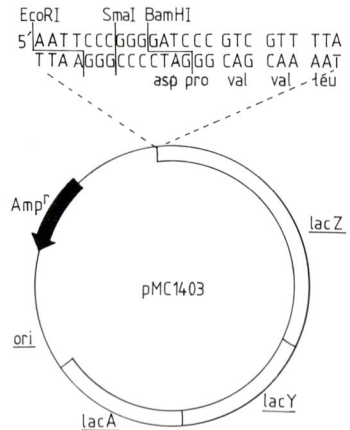

Figure 19   Fusion vector pMC1403. Plasmid pMC1403 is used for constructing *lacZ* gene fusions. The *lacZ* gene lacks its normal promoter, translational initiation codon and the first seven amino acid codons. DNA fragments cloned into the *Eco*RI, *Sma*I or *Bam*HI sites form gene fusions if they contain a promoter and translation initiation signals in the same reading frame as the truncated *lacZ* coding sequence. Construction of this, and similar plasmids, is described in Casadaban *et al.* (1983).

ing fragments from restriction digests of large DNA molecules. DNA fragments with promoter activity will produce blue colonies on X-gal (5-bromo-4-chloro-3-indolyl-β-D-galactoside) plates (Casadaban *et al.*, 1983). In addition, the effects of chromosomal regulatory mutations on the expression of β-galactosidase can be readily assayed. Vectors based on β-galactosidase can also be used to select chromosomal regulatory mutations or mutations in the promoter–operator sequences using chromogenic indicator plates. The many simple selections and screens for identifying colonies with mutations altering β-galactosidase expression have been described (Miller, 1972; Casadaban *et al.*, 1983).

Operon fusions only reflect transcriptional signals and not translational signals. This limitation can be overcome by constructing gene fusions which reflect both transcription and translation. The vector pMC1403 (Fig. 19) and similar derivatives have been constructed to allow the isolation of gene fusions such that both transcription and translation of β-galactosidase are dependent on inserted DNA (Casadaban *et al.*, 1980, 1983). The major disadvantage of these vectors is that the inserted fragment must have an appropriate restriction site in the correct reading frame to produce a hybrid protein. Specific vectors providing a means of overcoming this problem have recently been devised (Berman and Jackson, 1984). When a restriction fragment is inserted in the wrong reading frame, *lacZ* is not expressed (although translational reading frame slippage may give low level expression in high copy number plasmids). However, selection for Lac$^+$ derivatives of the plasmid almost invariably results in deletions which fuse the upstream gene to *lacZ* and produce a hybrid protein. Point mutations are rare because the Lac$^+$ phenotype requires that both *lacZ* and *lacY* are expressed. The 3' end of the deletion must be within the first 26 codons of *lacZ* in order to produce an active product. However, the 5' end of the deletions can be at any point within the inserted upstream gene which puts the fusion in the correct reading frame.

2   *Vectors based on galactokinase*

The pK0 series of vectors, originally developed by McKenney *et al.* (1981), has been extensively used to study both promoter and terminator function. The plasmids are similar to the *lacZ* plasmids and have been constructed specifically to detect transcriptional signals. The major difference between these vectors and the *lacZ* vectors, apart from the use of galactokinase as the easily assayable gene product, is that translation termination signals have been cloned in front of the *galK*

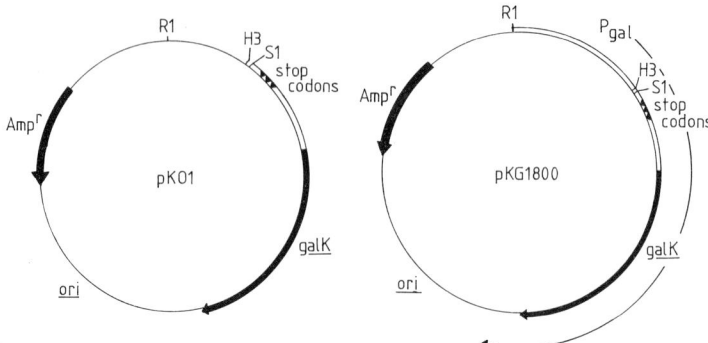

Figure 20 The two basic galactokinase vectors for studying promoters (pK01) and terminators (pKG1800) are illustrated. Each plasmid contains an ampicillin resistance gene (Amp^r) and the ColE1 origin of replication (*ori*). The *galK* gene is preceded by stop (translation termination) codons in each of the three reading frames (▲). pK01 has no promoter; *galK* is expressed only when promoters are inserted into the *Hin*dIII or *Sma*I sites. pKG1800 has the *gal* promoter directing transcription of *galK*. Terminators cloned in the *Hin*dIII or *Sma*I sites will reduce *galK* expression from this promoter. Restriction enzyme sites: R1 ≡ *Eco*RI; H3 ≡ *Hin*dIII; S1 ≡ *Sma*I. The construction of these vectors is described by McKenney *et al.* (1981). Many variants on the plasmids have been constructed with different cloning sites in place of the *Hin*dIII and *Sma*I sites.

gene in all three reading frames (Fig. 20) This prevents ribosomes which initiate translation from signals on the inserted promoter fragment proceeding into the *galK* gene. *galK* expression is thus dependent only on transcriptional signals and not on translational signals introduced by fragments cloned upstream. Many different vectors, with different upstream cloning sites, are now available. As with β-galactosidase there are several easy selections and screens for promoter or regulatory mutations on indicator plates (McKenney *et al.*, 1981). The pK0 vectors have also been adapted for the study of terminators. If a promoter is provided upstream of *galK* so that the gene is expressed, terminators can be cloned between the promoter and *galK* and their effect on transcription assessed (see Fig. 20).

Despite their many uses, there are a number of problems associated with *in vitro* fusion vectors which must always be borne in mind.

(i) Plasmid copy number can vary between strains and can also depend on the inserted promoter fragment. For example, when strong promoters are cloned into the pK0 vectors, transcription proceeds through *galK* and into the plasmid origin of replication. This can dramatically reduce copy number. The copy number should always be measured, either by assaying an independently expressed gene on the same plasmid or by dot–blot hybridization techniques.

(ii) Multicopy promoters and/or operators can titrate out regulatory molecules and thus exhibit abnormal regulatory responses. To overcome this, methods have been devised to incorporate *in vitro galK* fusions into the chromosome in single copy using a λ phage derivative (McKenney *et al.*, 1981). Fusion plasmids can also be introduced into the chromosome in single copy using PolA strains in which the plasmid cannot replicate autonomously. Unexpectedly, however, a given fusion can be expressed very differently when transcribed from the chromosome rather than from a plasmid (after taking copy number into account). The reason for this is unclear but may be due either to differential DNA supercoiling or to sequences distant from a promoter which can affect its function.

(iii) Although *lacZ* or *galK* expression is dependent on the inserted promoter, and is regulated in a similar manner, it is *not* true to say that expression is a measure of promoter strength. The reason why different fusions to the same promoter often express the downstream gene to very different extents is unknown but is probably due to differences in the 5′ end of the mRNA which alter its stability or translation efficiency.

3   *Gene cartridges*

An alternative approach to cloning promoters into fusion vectors is the use of gene cartridges. The first such cartridges developed contained the *lacZ* and *lacY* genes on a restriction fragment which could readily be excised and cloned into an appropriate restriction site in any plasmid-encoded gene (Casadaban *et al.*, 1980). Cartridges lacking the *lacZ* promoter but containing translational initiation signals will produce operon fusions when cloned downstream of an external promoter. Cartridges lacking the promoter, translational initiation signals and the first few codons of *lacZ* will produce gene (protein) fusions. Gene cartridges encoding genes other than *lacZ*, such as chloramphenicol acetyl transferase, and which are bounded by a variety of different restriction enzyme sites are now available (e.g. Close and Rodriguez, 1982), facilitating the *in vitro* construction of gene and operon fusions.

C   Uses of fusion technology

Various examples of the uses of gene and operon fusions have been referred to throughout this chapter. The use of fusions to study bacterial gene expression is becoming so widespread that a comprehensive coverage will not even be attempted. Here I will simply outline the sort of information which can be gained by using fusions.

The expression of β-galactosidase from a given fusion is dependent on transcription from the promoter to which it is fused and is regulated in a manner identical with that of the promoter. However, it should be emphasized that the level of β-galactosidase expression does not necessarily reflect the strength of that promoter. Different fusions to the same gene can give very different levels of β-galactosidase expression (see for example Stern *et al.*, 1984b). The reasons for this are not clear but are probably a combination of transcriptional polarity and altered mRNA stability introduced into the hybrid mRNA by the fusion.

The main uses of gene and operon fusions are as follows: (i) to determine the environmental conditions and/or stimuli which affect expression of any particular gene; (ii) to determine whether known regulatory genes affect the expression of any gene under study (both these uses are illustrated in Jamieson and Higgins, 1984); (iii) to identify those genes regulated by a given pleiotropic regulatory mutation; (iv) to identify genes the expression of which is regulated by any given external stimulus (see for example Cairney *et al.*, 1985); (v) to facilitate the isolation and characterization of mutations in regulatory genes; (vi) to isolate promoter and/or operator mutations. The last two uses rely on the ability to detect rare events which alter β-galactosidase production. This is readily achieved using one of a battery of different chromogenic indicators which on agar plates can detect single mutant colonies amongst many thousand wild-type cells (e.g. MacConkey agar, lactose tetrazolium, ONPG, X-gal (see Miller, 1972)) or simply the ability to utilize lactose as sole carbon source. A comparison between gene and operon fusions can also be used to determine whether regulation of a particular gene is transcriptional or translational (see for example Simons and Kleckner, 1983). Finally, although not connected with the study of gene expression, hybrid proteins produced from gene fusions have been very useful in aiding the identification of membrane proteins and in the genetic dissection of the protein secretory pathway. The development of fusion technology thus provides a powerful approach to the study of many problems associated with bacterial gene expression.

VII   Conclusions

It is hoped that this chapter has illustrated the many ingenious mechanisms by which *E. coli* regulates the expression of its genes. Rapid progress is now being made in elucidating the many subtleties of these mechanisms and unexpected discoveries are emerging with some regularity. It should be obvious not only that many important and interesting questions are unresolved also but that the technology is now available to tackle these problems. This promises to be an exciting area of research in the next few years.

VIII   Acknowledgements

I wish to convey my apologies to the authors of the many deserving studies which I have been unable to cite and to thank numerous colleagues for their contributions through discussion.

The author is a Lister Institute Research Fellow.

IX   References

Achord, D. and Kennell, D. (1974). *J. molec. Biol.* **90,** 581–599.
Adhya, S., Gottesman, M. and de Crombrugghe, B. (1974). *Proc. natn. Acad. Sci. USA* **71,** 2534–2538.
Aiba, H., Adhya, S. and de Crombrugghe, B. (1981). *J. biol. Chem.* **256,** 11905–11910.
Anderson, P. and Roth, J. (1981). *Proc. natn. Acad. Sci. USA* **78,** 3113–3117.
Anderson, W. F., Ohlendorf, D. H., Takeda, Y. and Matthews, B. W. (1981). *Nature, Lond.* **290,** 754–758.
Anderson, J., Ptashne, M. and Harrison, S. C. (1984). *Proc. natn. Acad. Sci. USA* **81,** 1307–1311.
Apirion, D. (1973). *Molec. gen. Genet.* **122,** 313–322.
Baker, T. A., Howe, M. M. and Gross, C. A. (1983). *J. Bacteriol.* **156,** 970–974.
Barrett, E. L., Kwan, H. S. and Macy, J. (1984). *J. Bacteriol.* **158,** 972–977.
Barry, G., Squires, C. and Squires, C. L. (1980). *Proc. natn Acad. Sci. USA* **77,** 3331–3335.
Baughman, G. and Nomura, M. (1984). *Proc. natn. Acad. Sci. USA* **81,** 5389–5393.
Beckwith, J. (1981). *Cell* **23,** 307–308.
Belfort, M., Pedersen-Lane, J., West, D., Ehrenman, K., Maley, G., Chu, F. and Maley, F. (1985). *Cell* **41,** 375–382.
Berman, M. L. and Jackson, D. E. (1984). *J. Bacteriol.* **159,** 750–756.
Bertrand, K., Squires, C. and Yanofsky, C. (1976). *J. molec. Biol.* **103,** 319–337.
Blundell, M. and Kennell, D. (1974). *J. molec. Biol.* **83,** 143–161.
Bochner, B. R. and Ames, B. N. (1982). *Cell* **29,** 929–937.
Bochner, B. R., Lee, P. C., Wilson, S. W., Cutler, C. W. and Ames, B. N. (1984). *Cell* **37,** 225–232.
Bossi, L. and Roth, J. R. (1980). *Nature, Lond.* **286,** 123–127.

Bremer, E., Silhavy, T. J., Weisemann, J. M. and Weinstock, G. M. (1984). *J. Bacteriol.* **158**, 1084–1093.

Buck, M. and Ames, B. N. (1984). *Cell* **36**, 523–531.

Buck, M. and Griffiths, E. (1982). *Nucl. Acids Res.* **10**, 2609–2624.

Cairney, J., Booth, I. R. and Higgins, C. F. (1985). *J. Bacteriol.*, **164**, 1224–1232.

Calos, M. P. and Miller, J. H. (1980). *Cell* **20**, 579–595.

Cannistraro, V. J. and Kennell, D. (1979). *Nature, Lond.* **277**, 407–409.

Casadaban, M. J. (1976). *J. molec. Biol.* **104**, 541–555.

Casadaban, M. J. and Chou, J. (1984). *Proc. natn. Acad. Sci. USA* **81**, 535–539.

Casadaban, M. J. and Cohen, S. N. (1979). *Proc. natn. Acad. Sci. USA* **76**, 4530–4533.

Casadaban, M. J. and Cohen, S. N. (1980). *J. molec. Biol.* **138**, 179–207.

Casadaban, M. J., Chou, J. and Cohen, S. N. (1980) *J. Bacteriol.* **143**, 971–980.

Casadaban, M. J., Martinez-Arias, A., Shapira, S. K. and Chou, J. (1983). *Methods Enzymol.* **100**, 293–308.

Christensen, T., Johnsen, M., Fiil, N. P. and Friesen, J. D. (1984). *EMBO J.* **3**, 1609–1612.

Chu, F. K., Maley, G. F., Maley, F. and Belfort, M. (1984). *Proc. natn. Acad. Sci. USA* **81**, 3049–3053.

Close, T. J. and Rodriguez, R. L. (1982). *Gene* **20**, 305–316.

Coleman, J., Green, P. J. and Inouye, M. (1984). *Cell* **37**, 429–436.

Csonka, L. N., Howe, M. M., Ingraham, J. L., Pierson, L. S. and Turnbough, C. L. (1981). *J. Bacteriol.* **145**, 299–305.

Dean, D. and Nomura, M. (1980). *Proc. Natn. Acad. Sci. USA* **77**, 3590–3594.

DiNardo, S., Voelkel, K. A., Sternglanz, R., Reynolds, A. E. and Wright, A. (1982). *Cell* **31**, 43–51.

Dixon, R., Alvarez-Morales, A., Clements, J., Drummond, D., Merrick, M. and Postgate, J. R. (1983). *In* "Advances in Nitrogen Fixation Research" (C. Veeger and W. E. Newton, eds) pp. 635–642. Nijhoff–Junk, Gröningen.

Drew, H. R. and Travers, A. A. (1984). *Cell* **37**, 491–502.

Ebright, R. H., Cossart, P., Gicquel-Sanzey, B. and Beckwith, J. (1984). *Nature, Lond.* **311**, 232–235.

Eisenberg, S. P., Yarus, M. and Soll, L. (1979). *J. molec. Biol.* **135**, 111–126.

Faelen, M., Mergeay, M., Gerits, J., Toussaint, A. and Lefebvre, N. (1981). *J. Bacteriol.* **146**, 914–919.

Farnham, P. J., Greenblatt, J. and Platt, T. (1982). *Cell* **29**, 945–951.

Fisher, R. and Yanofsky, C. (1984). *Nucl. Acids Res.* **12**, 3295–3302.

Friedman, D. I., Schauer, A. T., Baumann, M. R., Baron, L. S. and Adhya, S. L. (1981). *Proc. natn. Acad. Sci. USA* **78**, 1115–1118.

von Gabain, A., Belasco, J. G., Schottel, J. L., Chang, A. C. Y. and Cohen, S. N. (1983). *Proc. natn. Acad. Sci. USA* **80**, 653–657.

Gegenheimer, P. and Apirion, D. (1981). *Microbiol. Rev.* **45**, 502–541.

Gemmill, R. M., Tripp, M., Friedman, S. B. and Calvo, J. M. (1984). *J. Bacteriol.* **158**, 948–953.

Ghosh, B. and Das, A. (1984). *Proc. natn. Acad. Sci. USA* **81**, 6305–6309.

Gibson, M. M., Price, M. and Higgins, C. F. (1984). *J. Bacteriol.* **160**, 122–130.

Gilman, M. Z. and Chamberlin, M. J. (1983). *Cell* **35**, 285–293.

Giphart-Gassler, M., Plasterk, R. H. A. and van de Putte, P. (1982). *Nature, Lond.* **297**, 339–342.

Gitelman, D. R. and Apirion, D. (1980). *Biochem. Biophys. Res. Commun.* **96**, 1063–1070.

Glover, S. W., Firman, K., Watson, G., Price, C. and Donaldson, S. (1983). *Molec. gen. Genet.* **190,** 65–69.

Greenblatt, J. and Li, J. (1981a). *Cell* **24,** 421–428.

Greenblatt, J. and Li, J. (1981b). *J. molec. Biol.* **147,** 11–23.

Grossman, A. D., Erikson, J. W. and Gross, C. A. (1984). *Cell* **38,** 385–390.

Guarneros, G., Montanez, C., Hernandez, T. and Court, D. (1982). *Proc. natn. Acad. Sci. USA* **79,** 238–242.

Hall, M. N., Gabay, J., Debarbouille, M. and Schwartz, M. (1981). *Nature, Lond.* **295,** 616–618.

Har-El. R., Silberstein, A., Kuhn, J. and Tal, M. (1979). *Molec. gen. Genet.* **173,** 135–144.

Hawley, D. K. and McClure, W. R. (1983). *Nucl. Acids Res.* **11,** 2237–2255.

Higgins, C. F. and Ames, G. F.-L. (1982). *Proc. natn. Acad. Sci. USA* **79,** 1083–1087.

von Hippel, P. H., Bear, D. G., Morgan, W. D. and McSwiggen, J. A. (1984). *Ann. Rev. Biochem.* **53,** 389–466.

Holmes, W. M., Platt, T. and Rosenberg, M. (1983). *Cell* **32,** 1029–1032.

Hughes, K. T. and Roth, J. R. (1984). *J. Bacteriol.* **159,** 130–137.

Iida, S., Meyer, J., Kennedy, K. E. and Arber, W. (1983). *EMBO J.* **1,** 1445–1453.

Ikemura, T. (1981). *J. molec. Biol.* **146,** 1–21.

Izant, J. G. and Weintraub, H. (1984). *Cell* **36,** 1007–1015.

Jamieson, D. J. and Higgins, C. F. (1984). *J. Bacteriol.* **160,** 131–136.

Janel, G., Michelson, U., Nishimura, S. and Kiersten, H. (1984). *EMBO J.* **3,** 1603–1608.

Jaurin, B. and Normark, S. (1983). *Cell* **32,** 809–816.

Jaurin, B., Grundstrom, T., Edlund, T. and Normark, S. (1981). *Nature, Lond.* **290,** 221–225.

Jinks-Robertson, S. and Nomura, M. (1981). *J. Bacteriol.* **145,** 1445–1447.

Johnson, W. C., Moran, C. P. and Losick, R. (1983). *Nature, Lond.* **302,** 800–804.

Johnston, H. M. and Roth, J. R. (1981). *J. molec. Biol.* **145,** 735–756.

Kaplan, R. and Apirion, D. (1974). *J. biol. Chem.* **249,** 149–151.

Kassavetis, G. A. and Chamberlin, M. J. (1981). *J. biol. Chem.* **256,** 2777–2786.

Kassavetis, G. A. and Geiduschek, E. P. (1984). *Proc. natn. Acad. Sci. USA* **81,** 5101–5105.

Kingston, R. E. and Chamberlin, M. J. (1981). *Cell* **27,** 523–531.

Kingston, R. E., Nierman, W. C. and Chamberlin, M. J. (1981). *J. biol. Chem.* **256,** 2787–2797.

Kleckner, N. (1981). *Ann. Rev. Genet.* **15,** 341–404.

Konigsberg, W. and Godson, G. N. (1983). *Proc. natn. Acad. Sci. USA* **80,** 687–691.

Krueger, J. and Walker, G. C. (1983). *Methods Enzymol.* **100,** 501–509.

Kruger, K., Grabowski, P. J., Zaug, A. J., Sands, J., Gottschling, D. E. and Cech, T. R. (1982). *Cell* **31,** 147–157.

Kung, H., Spears, C. and Weissbach, H. (1975). *J. biol. Chem.* **250,** 1556–1562.

Kuroki, K., Ishii, S., Kano, Y., Miyashita, T., Nishi, K. and Imamoto, F. (1982). *Molec. gen. Genet.* **185,** 369–371.

Lamond, A. I. and Travers, A. A. (1983). *Nature, Lond.* **305,** 248–250.

Landick, R., Vaughn, V., Lau, E. T., VanBogelen, R. A., Erickson, J. W. and Neidhardt, F. C. (1984). *Cell* **38,** 175–182.

Lee, F. and Yanofsky, C. (1977). *Proc. natn. Acad. Sci. USA* **74,** 4365–4369.

Losick, R. and Pero, J. (1981). *Cell* **25,** 582–584.

McKay, D. B. and Steitz, T. A. (1981). *Nature, Lond.* **290,** 744–749.

McKenney, K., Shimatake, H., Court, D., Schmeissner, U., Brady, C. and Rosenberg, M. (1981). *In* "Gene Amplification and Analysis" (J. C. Chrikjian and T. S. Papas, eds) vol. II, pp. 383–415. North-Holland, Amsterdam.

Miller, J. H. (1972). "Experiments in Molecular Genetics." Cold Spring Harbor Laboratory, Cold Spring Harbor, NY.

Miller, J. H., Reznikoff, W. S., Silverstone, A. E., Ippen, K., Signer, E. R. and Beckwith, J. R. (1970). *J. Bacteriol.* **104,** 1273–1279.

Miozzari, G. F. and Yanofsky, C. (1978). *J. Bacteriol.* **133,** 1457–1466.

Mitchell, D. H., Reznikoff, W. S. and Beckwith, J. R. (1975). *J. molec. Biol.* **93,** 331–350.

Mizuno, T., Chou, M.-Y. and Inouye, M. (1984). *Proc. natn. Acad. Sci. USA* **81,** 1966–1970.

Mukai, F. H. and Margolin, P. (1963). *Proc. natn. Acad. Sci. USA* **50,** 140–148.

Muller-Hill, B. and Kania, J. (1974). *Nature, Lond.* **249,** 561–562.

Nilsson, G., Belasco, J. G., Cohen, S. N. and von Gabain, A. (1984). *Nature, Lond.* **312,** 75–77.

Nomura, M., Yates, J. L., Dean, D. and Post, L. E. (1980). *Proc. natn. Acad. Sci. USA* **77,** 7084–7088.

Nomura, M., Gourse, R. and Baughman, G. (1984). *Ann. Rev. Biochem.* **53,** 75–117.

Olson, E. R., Flamm, E. L. and Friedman, D. I. (1982). *Cell* **31,** 61–70.

Oppenheim, D. S. and Yanofsky, C. (1980). *Genetics* **95,** 785–795.

Oxender, D. L., Zurawski, G. and Yanofsky, C. (1979). *Proc. natn. Acad. Sci. USA* **76,** 5524–5528.

Pabo, C. O. and Lewis, M. (1982). *Nature, Lond.* **298,** 443–447.

Plasterk, R. H. A., Brinkman, A. and van de Putte, P. (1983). *Proc. natn. Acad. Sci. USA* **80,** 5355–5358.

Plasterk, R. H. A., Vollering, M., Brinkman, A. and van de Putte, P. (1984). *Cell* **36,** 189–196.

Putney, S. D. and Schimmel, P. (1981). *Nature, Lond.* **291,** 632–635.

Queen, C. and Rosenberg, M. (1981). *Cell* **25,** 241–249.

Reynolds, A. E., Felton, J. and Wright, A. (1981). *Nature, Lond.* **293,** 625–629.

Richardson, S. M. H., Higgins, C. F. and Lilley, D. M. J. (1984). *EMBO J.* **3,** 1745–1752.

Robertson, H. (1982). *Cell* **30,** 669–672.

Rosenberg, M. and Schmeissner, U. (1982). *In* "Interaction of Translational and Transcriptional Controls in the Regulation of Gene Expression" (M. Grunberg-Manago and B. Safer, eds). Elsevier, Amsterdam.

Rosenberg, M., Court, D., Shimatake, H., Brady, C. and Wulff, D. L. (1978). *Nature, Lond.* **272,** 414–423.

Roth, J. R. and Schmid, M. B. (1981). *In* "Stadler Symposium 13." University of Missouri, Columbia, MO.

Ryan, T. and Chamberlin, M. J. (1983). *J. biol. Chem.* **258,** 4690–4693.

Saito, H. and Richardson, C. C. (1981). *Cell* **27,** 533–542.

Sauer, R. T., Yocum, R. R., Doolittle, R. F., Lewis, M. and Pabo, C. O. (1982). *Nature, Lond.* **298,** 447–451.

Schlessinger, D., Jacobs, K. A., Gupta, R. S., Kano, Y. and Imamoto, F. (1977). *J. molec. Biol.* **110,** 421–439.

Schneider, E., Blundell, M. and Kennell, D. (1978). *Molec. gen. Genet.* **160,** 121–129.

Schumperli, D., McKenney, K., Sobieski, D. A. and Rosenberg, M. (1982). *Cell* **30,** 865–871.

Schwartz, M., Roa, M. and Debarbouille, M. (1981). *Proc. natn. Acad. Sci. USA* **78,** 2937–2941.

Shen, V., Cynamon, M., Daugherty, B., Kung, H.-F. and Schlessinger, D. (1981). *J. biol. Chem.* **256,** 1896–1902.

Shine, J. and Dalgarno, L. (1974). *Proc. natn. Acad. Sci. USA* **71,** 1342–1346.

Shuman, H. A. and Silhavy, T. J. (1981). *J. biol. Chem.* **256,** 560–562.

Siebenlist, U., Simpson, R. B. and Gilbert, W. (1980). *Cell* **20,** 269–281.

Silhavy, T. J., Berman, M. L. and Enquist, L. W. (1984). "Experiments with Gene Fusions." Cold Spring Harbor Laboratory, Cold Spring Harbor, NY.

Silverman, M. and Simon, M. (1980). *Cell* **19,** 845–854.

Silverman, M., Zieg, J., Hilmen, M. and Simon, M. (1979). *Proc. natn. Acad. Sci. USA* **76,** 391–395.

Simons, R. W. and Kleckner, N. (1983). *Cell* **34,** 683–691.

Sinden, R. R. and Pettijohn, D. E. (1981). *Proc. natn. Acad. Sci. USA* **78,** 224–228.

Singer, B. S., Gold, L., Shinedling, S. T., Colkitt, M., Hunter, L. R., Pribnow, D. and Nelson, M. A. (1981). *J. molec. Biol.* **149,** 405–432.

Spassky, A., Busby, S. and Buc, H. (1984). *EMBO J.* **3,** 43–50.

Steitz, T. A., Ohlendorf, D. H., McKay, D. B., Anderson, W. F. and Matthews, B. W. (1982). *Proc. natn. Acad. Sci. USA* **79,** 3097–3100.

Stern, M. J., Ames, G. F.-L., Smith, N. H., Robinson, C. and Higgins, C. F. (1984a). *Cell* **37,** 1015–1026.

Stern, M. J., Higgins, C. F. and Ames, G. F.-L. (1984b). *Molec. gen. Genet.* **195,** 219–227.

Sternglanz, R., DiNardo, S., Voelkel, K. A., Nishimura, Y., Hirota, Y., Becherer, K., Zumstein, L. and Wang, J. C. (1981). *Proc. natn. Acad. Sci. USA* **78,** 2747–2751.

Stokes, H. W. and Hall, B. G. (1984). *Proc. natn. Acad. Sci. USA* **81,** 6115–6119.

Stormo, G. D., Schneider, T. D. and Gold, L. M. (1982). *Nucl. Acids Res.* **10,** 2971–2996.

Travers, A. A., Lamond, A. I., Mace, H. A. F. and Berman, M. L. (1983). *Cell* **35,** 265–273.

Turnbough, C. L., Neill, R. J., Landsberg, R. and Ames, B. N. (1979). *J. biol. Chem.* **254,** 5111–5119.

Ullman, A. and Danchin, A. (1983). *Adv. Cyclic Nucl. Res.* **15,** 1–53.

Watson, N., Gurevitz, M., Ford, J. and Apirion, D. (1984). *J. molec. Biol.* **172,** 301–323.

Winkler, M. E. and Yanofsky, C. (1981). *Biochemistry* **20,** 3738–3744.

Wu, A. M., Chapman, A. B., Platt, T., Guarente, L. P. and Beckwith, J. (1980). *Cell* **19,** 829–836.

Wu, A. M., Christie, G. E. and Platt, T. (1981). *Proc. natn. Acad. Sci. USA* **78,** 2913–2917.

Yates, J. L. and Nomura, M. (1981). *Cell* **24,** 243–249.

Yates, J. L., Arfsten, A. E. and Nomura, M. (1980). *Proc. natn. Acad. Sci. USA* **77,** 1837–1841.

Youderian, P., Bouvier, S. and Susskind, M. M. (1982). *Cell* **30,** 843–853.

Zarucki-Schulz, T., Jerez, C., Goldberg, G., Kung, H.-F., Huang, K.-H., Brot, N. and Weissbach, H. (1979). *Proc. natn. Acad. Sci. USA* **76,** 6115–6119.

Zieg, J. and Simon, M. (1980). *Proc. natn. Acad. Sci. USA* **77,** 4196–4200.

Oncogenes

ALAN HALL

Chester Beatty Laboratories, Institute of Cancer Research, Fulham Road, London SW3 6JB, UK

I	Introduction.	61
II	The transformed cell	63
	A Density-dependent inhibition of growth	65
	B Anchorage-independent growth	65
	C Immortality	67
	D Growth factor requirements	68
	E *In vivo* tumorigenicity	68
	F Metastasis	69
III	RNA tumour viruses and viral oncogenes.	69
	A Rous sarcoma virus.	70
	B Other avian viruses.	72
	C Mammalian viruses.	74
IV	Cellular oncogenes	75
	A Proto-oncogenes	75
	B Proto-oncogenes and viral oncogenes	76
	C Viral activation of proto-oncogenes	79
	D Non-viral activation of proto-oncogenes	83
	E Biological assays to detect cellular oncogenes	92
V	Biochemistry of oncogene proteins	99
VI	Summary	108
VII	Acknowledgements	109
VIII	References	109

I Introduction

The cause of cancer has been a focus for scientific research for well over half a century but only during the last ten years has any real progress been made in understanding the genetic basis of the disease. The first oncogenic (cancer-inducing) virus was isolated over 70 years ago although we have only recently begun to appreciate how it causes cancer. This sudden leap forward has been the result of advances in many scientific disciplines but undoubtedly recombinant DNA technology has had a major impact. Through this new technology, coupled with improvements in tissue culture techniques, a tremendous amount has been learned in the last ten years about the genetics of the DNA and RNA tumour viruses and how they cause cancer in animals.

Until as recently as 1980 there was still much scepticism that studies of tumour viruses could lead to a greater understanding of the disease in humans and indeed it is still true that, as far as we know, the vast majority of human cancers are not caused by viruses. However, what has become apparent is that the knowledge accumulated during the 1970s about the mechanisms of action of tumour viruses has given real insight into the basis of human malignancy. The aim of this chapter is to show how these developments have occurred and have led to an enormous research effort in the 1980s to study the genetic basis of human cancer.

Clinically the most important aspect of human cancer is that it can kill the host. Experimentally, however, the transformation process that occurs when a normal cell becomes a cancer cell is not so clear cut. It is often the result of an accumulation of discrete changes in the morphology and biology of the cell, presumably as a consequence of separate genetic changes. This chapter will therefore begin with a discussion of some of the experimental markers that are available to study cellular transformation.

The common theme throughout the chapter will be oncogenes, i.e. genes which directly initiate and/or maintain the transformed phenotype of a cell. Oncogenes were first described as exogenous agents carried by tumour viruses, so-called viral oncogenes. Detailed analysis of RNA tumour viruses has been particularly informative and has drawn attention to the types of genes which can function as oncogenes, the biochemical processes that are important in maintaining the transformed phenotype of a cell and, perhaps more surprisingly, the genes which might be important in human cancer. The variety of mechanisms that RNA tumour viruses use to transform cells will therefore be discussed in section III and the first half of section IV.

The study of DNA tumour viruses has been extremely illuminating in the analysis of some of the individual steps leading to transformation. This group of viruses encompasses a much broader diversity of transforming agents than the RNA tumour viruses. At one extreme are members of the papova group of viruses, e.g. Simian virus 40 and polyoma, and the adenovirus group, which can readily transform susceptible cells in culture and can cause tumours in some animal hosts. At the other extreme are viruses such as Epstein–Barr virus (associated with human Burkitt's lymphoma) and the papilloma viruses (associated with a variety of human neoplasms), which appear to play some role in the transformation process but by themselves are not tumorigenic. The mechanisms by which the DNA tumour viruses initiate tumours will not be dealt with here (for reviews see Tooze, 1981, and Marshall and Rigby, 1984).

Although oncogenes were first described as the transforming agents carried by RNA tumour viruses, alterations in the structure and expression of some cellular genes have since been shown to be responsible for a number of the properties associated with cancer cells. As we shall see later, the discovery of these cellular oncogenes has caused a great deal of excitement and they are now under intense investigation as possible candidates for the genetic loci involved in human cancer. Recent techniques for detecting cellular oncogenes and for analysing their effects within cells will be described in section IV.

It is the protein products of oncogenes which mediate the transforming potential and the variety of biochemical activities associated with these proteins will be looked at in section V. Oncogene proteins somehow disrupt the normal growth control properties of a cell and an understanding of the biochemistry of this process is therefore essential.

By the end of this chapter I hope that the reader will have gained some appreciation of the excitement caused by the discovery of cellular oncogenes in human tumours. This has led to great anticipation in both the academic and the clinical communities that a breakthrough in the understanding and the eventual control of human cancer is at hand.

II The transformed cell

In 1911 Rous showed that an extract derived from a chicken fibrosarcoma could induce new tumours when injected into healthy animals (Rous, 1911). He was able to show that the extract contained a virus (subsequently called Rous sarcoma virus (RSV)) but it was not until 1956 that an *in vitro* assay for the virus was reported (Manaker and Groupe, 1956). The biological assays which have since been developed to observe the transformation process make use of the fact that tumour cells behave differently from normal cells both in *in vitro* culture and when injected into animals. Some of the differences are summarized in Table 1 and we shall now look at these in more detail.

Table 1 Some properties associated with normal and tumour cells and the ways in which these can be observed.

Normal cells	Tumour cells	Assay
Density-dependent growth	Density-independent growth	Focus formation on confluent monolayer
Anchorage dependent	Anchorage independent	Colony formation in soft agar
Serum dependent	Reduced serum dependence	Growth requirements in tissue culture
Limited lifespan	Immortal	Lifespan
Non-tumorigenic	Tumorigenic	Tumour formation in animals
Non-metastatic	Metastatic	Metastasis in animals

A   Density-dependent inhibition of growth

The extent to which normal cells will grow in culture is determined by their density. For example, fibroblasts, provided that they are supplied with the proper nutrients, will divide and grow in a culture dish but only until they have produced a confluent monolayer of cells. Other types of cells will often stop growing even before they reach confluence. The biochemical basis for this density-dependent inhibition of growth is not well understood. In the case of fibroblasts, there is some evidence that they need to flatten out over a surface before they can divide (Folkman and Moscona, 1978). Once the cells are in close contact they cannot spread out, and this leads to contact inhibition of growth. It has also been shown that some cells release growth-regulatory proteins which accumulate in the tissue culture medium to a level at which they inhibit growth (Hsu *et al.*, 1984).

In contrast, tumour cell growth is usually density independent; the cells continue to divide and pile up on top of each other (Fig. 1A). This forms the basis of the focus-forming assay. For example, if chicken fibroblasts are infected with RSV, the normal uninfected cells will grow but only until a monolayer is formed. Any cell which is infected and transformed by the virus will continue to divide, resulting in a clearly visible focus of piled-up cells. Cells of mesenchymal origin, e.g. fibroblasts, are particularly suited to this type of assay although it has been adapted for cells that normally grow in suspension, such as haematopoietic cells (Moscovici and Zanetti, 1970).

B   Anchorage-independent growth

Many normal cells must attach themselves to a solid support, such as a plastic surface, before they can divide and grow, whereas transformed cells will often grow in suspension in a semisolid medium such as soft agar (Macpherson and Montagnier, 1964) (Fig. 1B). As an example, if chicken fibroblasts infected with RSV are plated into soft agar the normal cells do not divide whereas the transformed cells form clearly visible colonies of growing cells (Weiss, 1970). This colony-forming assay measures the anchorage-independent phenotype of transformed cells and in general results in one of the best *in vitro* correlations with the *in vivo* tumorigenicity of cells. The assay can also be used to observe epithelial cell transformation.

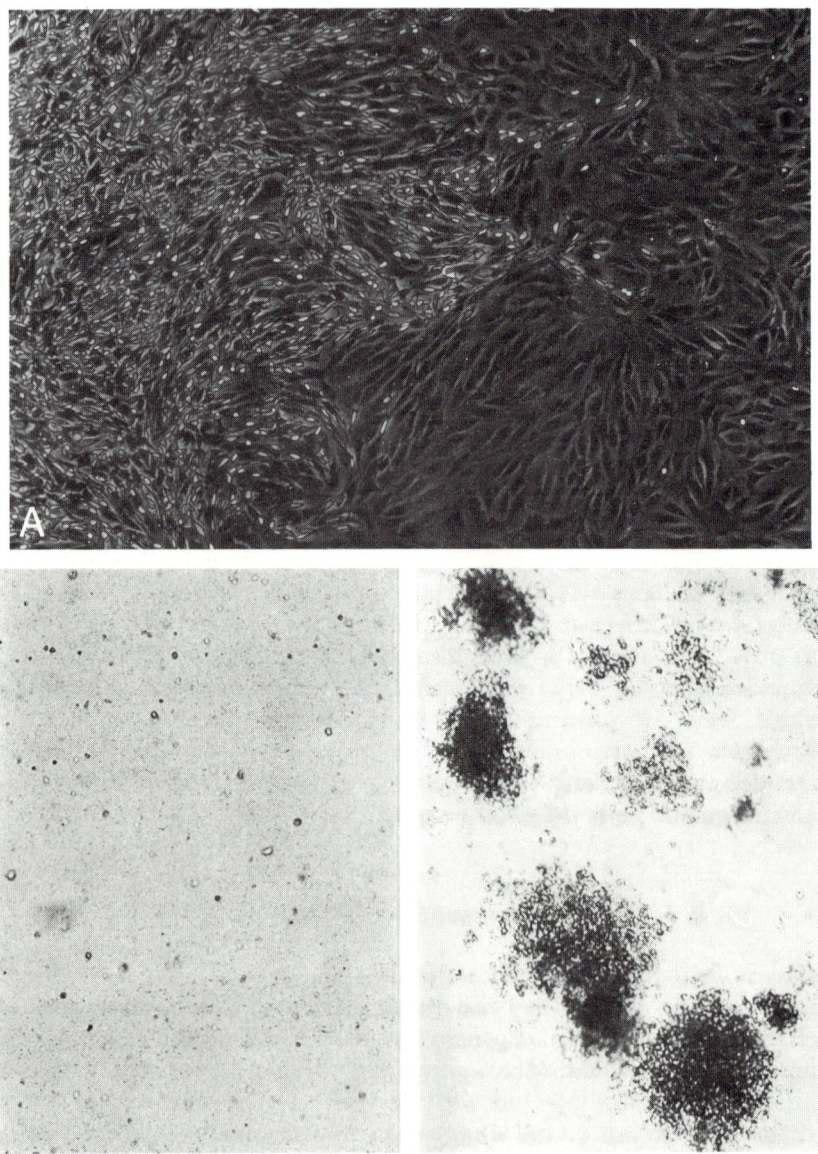

Figure 1 A, The border between a focus of transformed cells (left) and a confluent monolayer of normal fibroblasts (right): the transformed cells can be seen piling up on top of each other; morphologically they are smaller, more round and refractile to light. B, Normal (left) and transformed (right) fibroblasts seeded into soft agar. The normal cells do not divide and only single cells can be seen. Transformed cells do grow, giving rise to colonies.

C   Immortality

Normal cells have a limited lifespan when grown in culture. Human foetal fibroblasts, for instance, will undergo approximately 50 divisions before they senesce (stop growing) and die, whereas fibroblasts taken from adults undergo correspondingly fewer divisions (Hayflick and Moorhead, 1961). It appears that there is some internal programming for the cells to die and that they can somehow remember how many divisions they have undergone. However, immortal cell lines can be established from normal cells *in vitro* although the ease with which this can be achieved differs enormously between species and between different tissues. Syrian hamster fibroblasts, for example, can be immortalized by a single treatment with certain chemical carcinogens (Newbold *et al.*, 1982), whereas immortal human fibroblasts have never been obtained in this way.

Human tumour cells isolated from patients are immortal and it has been speculated that immortalization may be a prerequisite for cells to accumulate the necessary changes required for full tumorigenicity. Some normal cells, in particular stem cells, also seem to have at least an extended lifespan *in vivo* and it may be that these are the progenitor cells of human tumours. In this case "complete immortalization" could come later or may indeed be an artefact observed as a consequence of growing cells in culture. For example, the blood precursor cells have an extended and perhaps indefinite lifespan *in vivo* but they are normally continually differentiating into more mature cells which are committed to specific functions and which do have a limited lifespan. Human leukaemias may occur as a consequence of a block in the differentiation programme of the cells which leads to a massive build-up of immortal, poorly differentiated precursor cells. Limited lifespan and differentiation appear to be closely linked but they are distinct from the two previously described transformation phenotypes. This is best exemplified by the DNA tumour virus, polyoma. Polyoma has been shown to contain two oncogenes, one capable of immortalizing cells (that encoding large T antigen) and one producing the transformed phenotype (that encoding middle T antigen); both genes can produce their effects quite independently (Rassoulzadegan *et al.*, 1982). The genetic basis of immortality, or escape from senescence, in human tumours, however, is almost completely unknown.

D   Growth factor requirements

A variety of recipes is used to obtain optimal growth of different cells in culture. Normal fibroblasts will only grow well in defined medium (i.e. salts, sugars, amino acids and vitamins) if serum is added. Serum clearly contains factors essential for normal cell proliferation and many of these have been identified (Barnes and Sato, 1980). All cells require transferrin (an iron-binding protein) and insulin but other components such as growth factors (e.g. epidermal growth factor (EGF), platelet-derived growth factor (PDGF) and insulin-like growth factors (IGFs)), peptide and steroid hormones, cell attachment factors and small molecules such as lipids and trace metals are important to varying degrees.

Many tumour cells have a decreased dependence on exogenous factors and some will even grow in serum-free medium. The reasons for this are still not well understood; some no longer require any growth factors while others actually produce their own so-called transforming growth factors (TGFs). These were first observed in mouse fibroblasts transformed by the RNA tumour virus Moloney sarcoma virus (De Larco and Todaro, 1978). The virally transformed cells release mitogenic factors and one of these, TGFα, can bind to the EGF receptor and stimulate the cell's own growth (autocrine stimulation) as well as that of its neighbours (paracrine stimulation) (Sporn and Todaro, 1980). A second polypeptide, TGFβ, is also released from the transformed cells (Anzano *et al.*, 1983); this does not bind to the EGF receptor but acts synergistically with TGFα. Many human tumours have since been found to release TGFs (Todaro *et al.*, 1980) and their biological and biochemical properties are being studied.

E   *In vivo* tumorigenicity

It might be expected that the ultimate test for transformation would be to introduce the cells into an animal and to look for the appearance of a tumour; unfortunately this is often difficult to achieve. Even if an experimentally manipulated cell is reintroduced into a syngeneic animal, it might still be rejected by the immune system. The introduction of nude mice, a genetically determined athymic strain which cannot activate a T-cell response to foreign cells, has helped to overcome this problem. However, it is still true that the majority of human tumours, taken directly from patients, do not grow even in the nude mouse.

F   Metastasis

Clinically, one of the most important aspects of tumour cells is their ability to metastasize; i.e. cells dislodged from their original site are carried by body fluids to other sites, in particular the lung or lymph nodes, and then form secondary tumours. It is these secondary tumours which usually kill the patient. Metastasis in experimental systems is often difficult to evaluate as the tumours that grow at the site of inoculation will usually kill the animals before secondary tumours have time to form. To overcome this problem, two techniques have been developed: in the first, tumour cells are injected intravenously to see whether they will move to the lungs or other organs; in the second, cells are inoculated subcutaneously and allowed to form a tumour which is then surgically removed and the animal is allowed to recover to see whether metastases develop.

Table 1 lists the properties of normal and transformed cells that I have discussed. Not all tumours have all the characteristics shown. For instance, although primary cultures of fibroblasts can be morphologically transformed by RNA tumour viruses they are generally not immortalized (Ponten, 1970). Furthermore, the virally induced animal tumours which have been so useful experimentally in defining the properties of transformed cells are mainly sarcomas, i.e. tumours derived from cells of mesenchymal origin. The majority of human cancers are carcinomas, i.e. derived from cells of epithelial origin, and these do not necessarily behave similarly in *in vitro* assays.

It has long been thought that human cancer is the result of an accumulation of several genetic alterations (Cairns, 1978) and it is tempting to speculate that these are reflected in the discrete phenotypic changes observed experimentally. The type of alteration necessary may be dependent on the kind of cell affected or alternatively the same change may have different phenotypic consequences in different cells. It is looking increasingly likely that many of the alterations involve specific cellular oncogenes and it is hoped that, by identifying all possible oncogenes and analysing their effects in many cell types, a better understanding of the origin of the large array of human tumours will be obtained. In the next section we shall look at the most important source of oncogenes so far, the RNA tumour viruses.

III   RNA tumour viruses and viral oncogenes

The RNA tumour viruses are a closely related group of viruses that carry their genetic information in the form of a single-stranded RNA molecule. On entering a cell the viral RNA is first converted into a double-stranded DNA copy by the virally encoded enzyme reverse transcriptase (Baltimore, 1970; Temin and Mizutani, 1970); hence these viruses are often referred to as retroviruses. When a cell is transformed by an RNA tumour virus, at least one DNA copy of the virus is integrated into the host genome as a provirus (for a review of RNA tumour viruses see Weiss et al., 1985). One subgroup of this family of viruses will form tumours in animals within 2–4 weeks of infection; these acutely transforming viruses have been found to have evolved a common mechanism for achieving acute pathogenicity—the viral oncogene.

A   Rous sarcoma virus

As already mentioned, the RSV will induce fibrosarcomas when introduced into chickens and will transform chicken fibroblasts in vitro, as visualized by the focus-forming assay and the colony-forming assay. It will also transform epithelial and haematopoietic cells in vitro (Kahn et al., 1984). RSV carries three genes essential for replication and viral particle production (Fig. 2): gag (group-specific antigen), pol (the reverse transcriptase) and env (the viral envelope glycoprotein). A fourth gene, src, is present in the RSV genome and it is this gene that is responsible for the tumorigenic properties of the virus. An outline of how the virus is transcribed and translated is shown in Fig. 2.

The formal proof that the src region contained an oncogene came from a variety of genetic analyses. Toyoshima and Vogt (1969) and Martin (1970) isolated the first mutants of RSV that were temperature sensitive for transformation but not for growth. They would only transform cells at the permissive temperature, 37 °C, and not at the non-permissive temperature, 42 °C, but they grew equally well at both temperatures. This indicated that RSV contained a distinct gene responsible for its transforming ability. Since then many different mutants have been derived, induced either by physical (e.g. ultraviolet light) or by chemical (e.g. 5-bromodeoxyuridine) agents. Non-conditional transformation-defective mutants have also been isolated; these contain deletions or insertions within the viral genome and are much more stable (Bernstein et al., 1976). By analysing which nucleotides were missing in mutant genomes, the transforming gene, called src, was mapped to the 3' end of the virus (Wang et al., 1975; Coffin and Billeter, 1976). The complete nucleotide sequence of the RSV genome has since

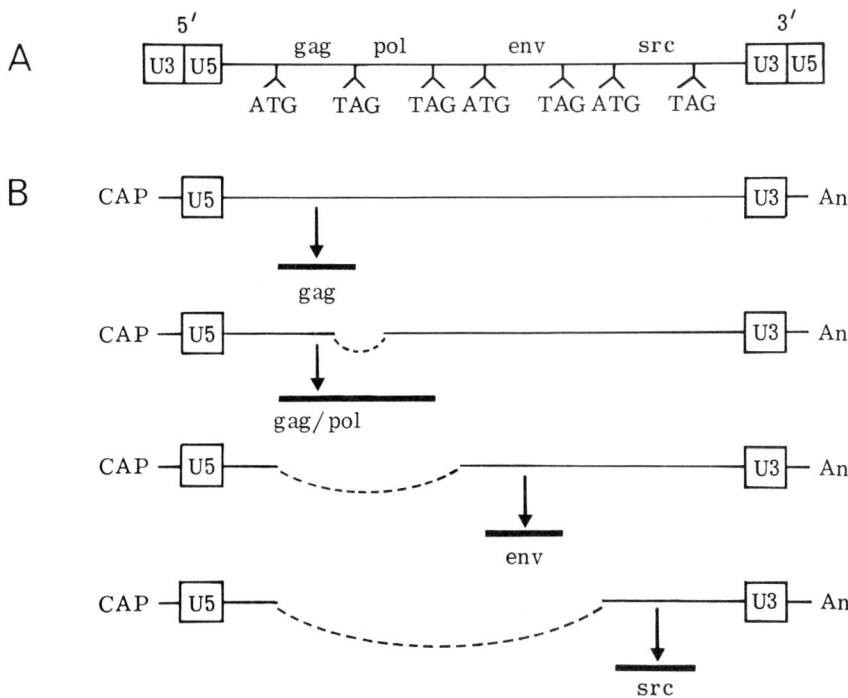

A

5'   3'

Figure 2 A, Genomic structure of a Rous sarcoma provirus. The provirus is approximately 9.3 kb long and is bounded by two long terminal repeats (LTRs) each about 350 bp long. The U3 block of sequence within the 5' LTR contains the promoter for transcription and U5 in the 3' LTR contains transcriptional stop signals. B, mRNAs (——) and proteins (━━) derived from transcription and translation of the virus. The RNAs are capped and polyadenylated. Each RNA contains a common 5' leader sequence but splicing (- - -) removes one or more termination codons so that all four protein products are made from different mRNA species.

been obtained (Schwartz *et al.*, 1983); the *src* region has an open-reading frame of 1590 nucleotides coding for a 526 amino acid protein with a molecular weight of approximately 60 000 daltons. This protein, designated pp60$^{v\text{-}src}$, is found in all RSV-transformed cells, although the transformed phenotypes elicited by *src* are varied and depend on the cells involved. The biochemical properties of the protein will be discussed in section V.

The early 1970s were a very exciting time for tumour virologists. The formal proof that a single gene carried by a retrovirus could cause cancer in animals stimulated a great deal of research, much of it centred on characterizing other RNA tumour viruses and defining their oncogenes. Unfortunately, unlike RSV, all other acutely transforming RNA viruses so far isolated are replication defective, i.e. they carry deletions and can only replicate in the presence of a helper virus which supplies the lost functions. The genetic analysis of these viruses is therefore complicated by the presence of helper-virus sequences. The use of recombinant DNA technology has overcome this problem and it is now possible to clone DNA copies of the viruses, to engineer defined mutations into the clone and then to reintroduce the DNA into cells to test for transforming ability. This procedure has enabled many new oncogenes to be identified and characterized.

B   Other avian viruses

Well over a dozen acutely transforming retroviruses with a range of pathogenic effects have been isolated from chickens (Table 2). Four other sarcomagenic viruses are known, each containing a different viral oncogene. Fujinami sarcoma virus appears to transform a very similar spectrum of cells to RSV but contains a distinct oncogene, *v-fps* (Hanafusa *et al.*, 1980). The Y73 virus isolate has been shown to contain the *v-yes* oncogene (Kitamura *et al.*, 1982), UR2 virus contains *v-ros* (Neckaneyer and Wang, 1984) and SK770 virus contains *v-ski* (Stavnezer

Table 2   Viral oncogenes and the origin of the first isolate.

Avian		Mammalian	
Oncogene	Origin	Oncogene	Origin
v-src	Chicken	*v-mos*	Mouse
v-fps (fes)	Chicken	*v-fos*	Mouse
v-yes	Chicken	*v-abl*	Mouse
v-ros	Chicken	*v-raf (mil)*	Mouse
v-ski	Chicken	*v-Ha-ras*	Rat
v-erbA	Chicken	*v-Ki-ras*	Rat
v-erbB	Chicken	*v-fes (fps)*	Cat
v-myc	Chicken	*v-fms*	Cat
v-myb	Chicken	*v-fgr*	Cat
v-ets	Chicken	*v-sis*	Monkey
v-mil (raf)	Chicken		
v-rel	Turkey		

et al., 1981); although all these viruses produce sarcomas in chickens they each contain a different oncogene.

The study of the defective avian viruses that can induce leukaemia has uncovered some particularly interesting oncogenes. Several independently isolated viruses contain the *v-myc* oncogene. The prototype virus MC29 will produce a variety of tumours *in vivo* (myelocytomatosis, endotheliomas and kidney and liver carcinomas) and *in vitro* will transform fibroblasts and macrophages (Graf and Beug, 1978). Mutants have been isolated that have a reduced capacity to transform macrophages whilst retaining their ability to transform fibroblasts (Ramsey *et al.*, 1980). These mutants have altered v-myc protein products, and such an analysis is helping to dissect functionally distinct domains on the protein. MC29 has an unusually long (greater than two months) latent period for tumour formation; also, although primary chick fibroblasts transformed by the virus are anchorage independent and not contact inhibited, they are not immortal and more importantly they do not form tumours in nude mice (Royer-Pokora *et al.*, 1978). This has led to speculation that the *myc* oncogene itself does not produce the full transformed phenotype but enables cells to proliferate abnormally until some kind of second event occurs resulting in true tumour cells (Palmieri *et al.*, 1983).

The MH2 virus isolate contains in addition to *v-myc* a second oncogene, *v-mil* (Coll *et al.*, 1983), also called *v-mht* (Kan *et al.*, 1984). It might be imagined that two oncogenes could cooperate with each other, perhaps to extend the host range or to alter the pathogenicity of a virus. In the case of MH2 there is some evidence that it is indeed more oncogenic *in vivo* than the MC29 isolate (Linial, 1982; Bechade *et al.*, 1985). Other data suggest that *v-mil* alone can act as an oncogene (Rapp *et al.*, 1983; Kan *et al.*, 1984).

Two viral isolates that contain the *v-erbB* gene have been characterized. Avian erythroblastosis virus (AEV) contains two genes, *v-erbA* and *v-erbB* (Lai *et al.*, 1979), whereas the AEV-H isolate contains only the *v-erbB* sequences (Hihara *et al.*, 1983). In the case of AEV, deletions have been made in one or other of the genes to generate A^-B^+ and A^+B^- viruses (Frykberg *et al.*, 1983) and it was found that the *v-erbB* gene is essential for transformation of erythroblasts and fibroblasts. The presence of *v-erbA* in the virus, however, does result in an altered spectrum of disease, apparently by blocking cell differentiation early within the erythroid lineage. *v-erbA* has been sequenced and it has some homology to carbonic anhydrase, although the significance of this is not known (Debuire *et al.*, 1984). It appears that *v-erbA* cannot function as an oncogene but can potentiate the oncogenic effects of *v-erbB*.

The oncogene *v-myb* has been identified in two viral isolates, avian myeloblastosis virus (AMV) and E26, and has since been sequenced (Rushlow *et al.*, 1982). AMV induces purely myeloid leukaemias, whereas E26 induces leukaemia mainly of erythroid origin (Radke *et al.*, 1982). Only the E26 isolate can transform fibroblasts *in vitro* although, as with *v-myc*-transformed fibroblasts, these are not tumorigenic in nude mice (Palmieri *et al.*, 1983). The major genetic difference between the two viruses is that, in addition to *v-myb*, E26 contains a second gene *v-ets* (Leprince *et al.*, 1983). The presence of *v-ets* clearly alters the preference of the virus for cell type although it is not known whether it can act as an oncogene in its own right.

In all, more than ten viral oncogenes have been identified in viruses isolated from chickens and one, *v-rel*, from a turkey reticuloendotheliosis virus (see Table 2).

C   Mammalian viruses

In 1964, whilst passaging Moloney murine leukaemia virus (Mo-MLV) through rats, Harvey obtained an isolate which when reintroduced into rats would induce sarcomas within 2–4 weeks (Harvey, 1964). Harvey murine sarcoma virus was the first acutely transforming mammalian retrovirus isolated. In similar experiments, carried out somewhat later, Kirsten murine leukaemia virus was passaged through rats and another highly sarcomagenic virus, now called Kirsten murine sarcoma virus, was isolated (Kirsten and Mayer, 1967). Both of these highly oncogenic isolates were derived from a virus which did not carry an oncogene, but somehow they acquired one whilst being passaged in the rat. The new genes present in these two viruses were found to be closely related to each other although not identical and have been called *v-Ha-ras* and *v-Ki-ras* (Ellis *et al.*, 1981). As a consequence of acquiring an oncogene, the virus had to lose many of its replicative functions (only 20% of the original murine leukaemia virus remains); hence, like all other RNA tumour viruses (except RSV) the *ras*-containing viruses are defective.

Two acutely transforming viruses have been independently isolated from mice infected with Mo-MLV. One, the Abelson murine leukaemia virus, contains the *v-abl* oncogene (Reddy *et al.*, 1983) and the other, Moloney murine sarcoma virus, contains *v-mos* (Van Beveren *et al.*, 1981). The FBJ murine osteosarcoma virus was isolated from a spontaneously occurring murine osteosarcoma and contains the *v-fos* oncogene (Curran *et al.*, 1982). The murine sarcoma virus MSV3611 contains the single oncogene *v-raf*, which has been shown to be the murine equivalent of the chicken *v-mil* sequence identified in the MH2 virus (Jansen *et al.*, 1984).

Several feline sarcoma viruses (FeSVs) have been isolated from cats infected with feline leukaemia virus. ST-FeSV and GA-FeSV contain an oncogene called *v-fes*, the SM-FeSV isolate contains the *v-fms* oncogene and GR-FeSV contains *v-fgr*. These oncogenes have been sequenced and *v-fes* has been shown to be a feline equivalent of the chicken *v-fps* gene described earlier (Hampe *et al.*, 1982; Naharro *et al.*, 1984). The simian sarcoma virus is the only acutely transforming retrovirus isolated from a primate species, the pet woolly monkey. The virus has been sequenced and shown to contain an oncogene *v-sis* (Sushilkumar *et al.*, 1983).

It should be apparent from this section that the acutely transforming RNA tumour viruses have been an invaluable source of oncogenes (see Table 2). Although the majority of these viruses induce the formation of sarcomas, some do transform haematopoietic cells and one group, the *myc*-containing viruses, can give rise to carcinomas. Furthermore, the viruses that contain more than one oncogene should be very useful in analysing cooperation between different transforming agents. It is widely thought that the number of oncogenes that can be identified by analysing retroviruses is reaching a plateau. Many of the oncogenes described have been identified several times in viruses isolated from the same and from different species and it seems unlikely that many more will be found.

After this introduction to the biology of the transformed cell and the genetics of the RNA tumour viruses, we are now in a position to move on from viral oncogenes and to look at the discovery and subsequent analysis of cellular oncogenes.

IV   Cellular oncogenes

A   Proto-oncogenes

As early as 1973 it was postulated that the oncogenes acquired by retroviruses might be derived from cellular genes present in the host (Weiss, 1973). In 1976 this hypothesis was confirmed when it was shown that cDNA specific for the *v-src* region of RSV could detect closely related sequences in the genome of normal chicken cells (Stehelin *et al.*, 1976). These sequences correspond to a gene, now called cellular *src* or *c-src*, which is present at one copy per haploid genome and which has subsequently been found in all other vertebrate species including man (Spector *et al.*, 1978).

Since the discovery of cellular sequences homologous to *v-src*, cellular counterparts for each of the viral oncogenes listed in Table 2 have also been found in all vertebrate species (Bishop, 1983). These cellular sequences are known as cellular oncogenes or proto-oncogenes, and each of the chicken, mouse, cat and monkey tumour viruses discussed earlier had picked up genetic information from proto-oncogenes present in the host. It is still difficult to imagine what, if any, the selective pressures were for the evolution of the oncogenic viruses in natural populations but, whatever the explanation, it must account for the fact that a few have picked up two cellular sequences in, presumably, two independent steps. From these studies it is clear that within the vertebrate genome there are at least 20 genes which we know under certain circumstances (i.e. in a retrovirus) can cause cancer. The possibility must be considered that these proto-oncogenes could be converted into *bona fide* cellular oncogenes and that this process might be involved in the development of human cancer.

The expression of proto-oncogenes has been detected in many normal cells. The *c-ras* and *c-myc* genes are transcribed in almost all cells (although the levels are low at about 5–20 molecules of RNA per cell) whereas most others appear to be more tissue specific. For example, *c-myb* is expressed in many cells of the haematopoietic system but not elsewhere (Westin *et al.*, 1982a, 1982b) and *c-sis* RNA has been detected in very few normal cell types, including rapidly dividing cells of the human placenta and endothelial cells (Barrett *et al.*, 1984; Collins *et al.*, 1985). The expression of some cellular oncogenes does not appear to have any obvious relationship to the role of the genes in oncogenesis. For example, *c-src* expression seems to be high in adult chicken brain, which is a non-dividing tissue (Barnekow and Bauer, 1984). More recently it has been shown that the rat central nervous system contains up to twentyfold higher levels of c-src protein than fibroblasts (Brugge *et al.*, 1985). Coupled with the fact that the *c-onc* sequences are so conserved between species (some have been detected in *Drosophila* and yeast (Shilo and Weinberg, 1981; Defeo-Jones *et al.*, 1983)) it seems likely that the gene products play an essential and basic role in normal cellular growth and development.

B   Proto-oncogenes and viral oncogenes

It is important to ask why proto-oncogenes expressed in cells do not lead to transformation, whereas viral oncogene expression clearly does. Two explanations have been proposed. The first is that there is a quantitative difference in the level of expression of the cellular and viral genes, the former being transcribed from their own genomic promoters, the latter from viral promoters. This could also be linked to inappropriate expression of the gene in a particular cell type. The second possibility is that the protein products are qualitatively different and that amino acid sequence differences are biologically important. We shall now look at some examples of the ways in which viral oncogenes differ from proto-oncogenes.

1   *src*

c-src has been molecularly cloned from both chicken and human DNA and nucleotide sequencing has revealed just how similar the protein coding regions are to *v-src* (Takeya and Hanafusa, 1982). The viral and cellular genes code for proteins with molecular weights of approximately 60 000 daltons (*v-src*, 526 amino acids; *c-src*, 533 amino acids). They differ at eight internal amino acids and in addition the 19 carboxy-terminal residues of chicken *c-src* are replaced by a new set of 12 residues in *v-src*. Unlike *v-src*, however, the *c-src* gene is a very complex structure with 11 introns. The exact mechanism by which RSV acquired genomic *c-src* information is still not clear (Wyke, 1983); it is possible that, during a round of infection, a non-oncogenic progenitor of RSV transduced genomic DNA after viral integration and excision and then the introns were removed by processing, or alternatively it may have somehow incorporated *c-src* mRNA.

The differences between *c-src* and *v-src* have recently been addressed by making use of *in vitro* recombinants of the viral and cellular genes, i.e. by replacing different parts of the cloned chicken *c-src* gene by sequences from *v-src* (Hanafusa *et al.*, 1984; Shalloway *et al.*, 1984). These recombinants were then reintroduced into cells using an appropriate expression vector and it was shown that some of the amino acid changes present in *v-src* were biologically important for transformation. In agreement with this, high level expression of *c-src* alone does not transform cells (Parker *et al.*, 1984). However, using controllable expression vectors, it has also been shown that if *v-src* is expressed in cells at levels comparable with that at which *c-src* is expressed in normal cells then no transformation is observed (Jakobovits *et al.*, 1984). It may be that both qualitative and quantitative changes in *src* expression are required for transformation.

2   *mos*

The cellular and viral *mos* genes have both been sequenced and have been found to differ by 11 amino acids (Van Beveren *et al.*, 1981). When the murine *c-mos* gene is linked to a strong promoter this construct can transform cells *in vitro* (Blair *et al.*, 1981) but high levels of *c-mos* RNA and protein are required. In contrast, only one to ten copies of *v-mos* RNA per cell are necessary for transformation (Wood *et al.*, 1983) and it seems likely that qualitative differences between the viral and cellular *mos* sequences are important.

3   *fos*

The cellular and viral *fos* genes have an interesting relationship to each other. The first 332 amino acids of *v-fos* and murine *c-fos* differ in only five positions but the remaining 49 amino acids are completely different. It appears that 104 bp at the C terminus of *c-fos* have been deleted in *v-fos* and, although this changes the reading frame and alters all subsequent amino acids, the overall lengths of the proteins are almost identical (Van Beveren *et al.*, 1983, 1984). The *c-fos* gene will transform fibroblasts in culture provided that a transcriptional enhancer element is added and if the 3' non-coding sequences are removed. It is thought that the 3' sequences inhibit the translation of *c-fos* RNA and these sequences are missing in the virus. There is no evidence that the amino acid differences are important.

4   *ras*

One of the best studied comparisons between proto-oncogenes and viral oncogenes is that for the *ras* genes. The *c-ras* genes can be converted into transforming genes by linking them to strong promoters (Chang *et al.*, 1982a). However, it is the activation of *c-ras* genes in human tumours by amino acid substitutions that has caused the greatest interest and this will be dealt with in more detail later (see Section IV E).

5   *myc*

The product of the MC29 virus is a fusion protein between sequences derived from the *gag* and *myc* genes. It is not known whether the gag protein sequences have any influence on the biological properties of myc. The introduction of amino acid alterations into the v-myc portion of the protein can affect the target cell specificity for transformation by the MC29 virus (Ramsey *et al.*, 1980). Since c-myc and v-myc differ in seven amino acids (Watson *et al.*, 1983), it is possible that these are functionally important changes although this has not been proved. As we shall see later, there is also some suggestion that the 5′ untranslated region of *c-myc* mRNA has a regulatory role in translation (Leder *et al.*, 1983; Saito *et al.*, 1983) and this region is not present in the MC29 virus.

6 *erbB*

The *c-erbB* gene has recently been shown to be the EGF receptor gene (Downward *et al.*, 1984). *v-erbB* is a truncated form of the gene which yields a protein about half the size of the cellular receptor protein. This altered protein product is unable to bind EGF and presumably provides a constitutive signal for the cells to divide.

7 *sis*

c-*sis* has been shown to be the gene for the B chain of PDGF (Waterfield *et al.*, 1983; Johnsson *et al.*, 1984). *v-sis* has an almost identical sequence although the virus actually produces a fusion protein between its env protein and the v-sis sequences, the env portion being important for export of the protein. It is thought that inappropriate expression might be the major difference in this case.

C Viral activation of proto-oncogenes

Not all RNA tumour viruses contain viral oncogenes, yet they undoubtedly cause cancer in susceptible hosts. In such cases the latent period is generally of the order of several months as opposed to several weeks for the acutely transforming variety. The mechanism by which the non-acutely transforming viruses cause cancer has given further insight into the role of cellular oncogenes in neoplastic transformation and has also led to the identification of new oncogenes. We shall now look at three groups of viruses that have been, or are being, closely analysed.

1   *Avian leukosis virus*

Avian leukosis virus (ALV) causes B-cell lymphomas in chickens. After infection of chicken cells *in vitro* the virus is found integrated at many positions in the host DNA, but in DNA isolated from tumours from infected chickens the provirus is always found at the same locus. In 1981 it was shown that this locus was the chicken *c-myc* gene (Hayward *et al.*, 1981). Normal cells contain about five copies of *c-myc* RNA but tumour cells contain fiftyfold more and on closer examination it was found that the avian leukosis provirus was integrated very close to the gene. In many cases most of the virus is deleted but an LTR containing an enhancer and a promoter sequence is always present. *c-myc* is an unusual gene: it contains three exons and although exon I is not translated into protein it is partially conserved between species. It has been postulated that it may have some regulatory role (Leder *et al.*, 1983; Saito *et al.*, 1983) and most integrations of ALV result in the separation of exons II and III from the normal promoter and exon I (Fig. 3) (Payne *et*

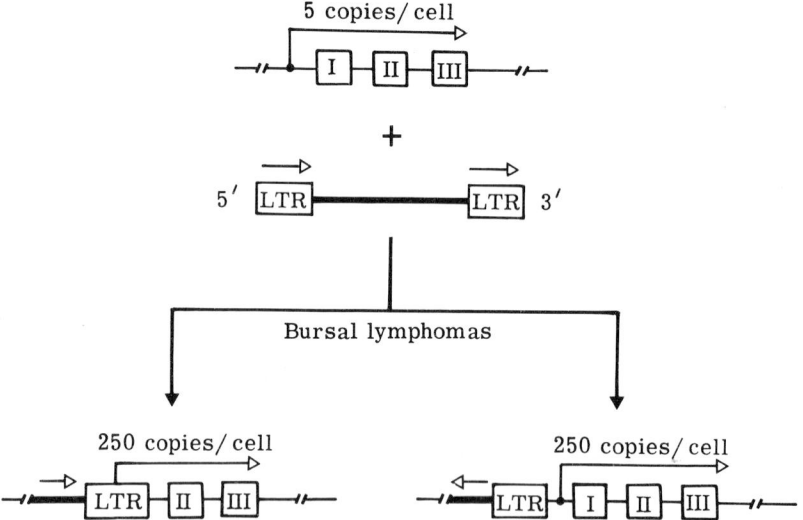

Figure 3   Effect of ALV integration on transcription of the chicken *c-myc* gene. *c-myc* consists of three exons: exon I is a 5′ untranslated sequence; exons II and III code for the myc protein. In most cases ALV integrates between exons I and II so that the normal promoter is replaced with a viral LTR promoter. In a few cases the virus is found in the opposite orientation and an enhancer element in the viral LTR probably increases transcription from the normal *c-myc* promoter.

al., 1982; Shih *et al.*, 1984). In the majority of cases the orientation of the LTR promoter is in the same transcriptional sense as the *c-myc* gene and the consequence of the integration event is to place exons II and III under the control of a viral promoter. In a minority of tumours, the transcriptional sense is opposite to that of *c-myc* and in one case analysed the provirus is actually 3′ of *c-myc* (Payne *et al.*, 1982). In these cases it is thought that the enhancer element in the LTR increases transcription from the normal genomic promoter. The mechanism of activation of *c-myc* by ALV is therefore complex and depends on the site of integration. The result, however, is an increase in transcription usually coupled to the removal of the 5′ untranslated sequences which have a putative regulatory role.

The activation of *c-myc* is only one step in the development of B-cell lymphomas induced by ALV and subsequent changes, presumably in other cellular oncogenes, are required. One of these has already been identified (Goubin *et al.*, 1983; see also section IV). It is interesting to note that activation of *c-myc* leads to B-cell lymphomas, whereas MC29 virus, containing *v-myc*, gives rise to carcinomas and sarcomas but not B-cell lymphomas. It is possible that although the non-oncogene sequences in MC29 do not contribute directly to tumorigenesis they may set limits on the cell type which can be subjected to the effects of *v-myc*. The expression of the gene is after all under the control of viral LTR promoter sequences. Alternatively, as we have already seen, there is some evidence that amino acid sequence differences between the viral and cellular oncogenes might alter the target cell specificity. More recently, avian leukosis proviruses have been found integrated at the *c-erbB* locus in chicken erythroblastosis (Fung *et al.*, 1983). This results in an elevated level of *c-erbB* transcripts.

The experiments of Hayward *et al.* (1981) were a milestone in many ways. They were the first demonstration that proto-oncogenes could be activated other than by being transduced by a retrovirus, removing any doubt that these sequences are indeed potential endogenous oncogenes.

2 *Mouse mammary tumour virus*

It was known as early as 1933 that mammary adenocarcinomas of some strains of mice could be maternally inherited via an extrachromosomal element. This element is now known to be a retrovirus, the mouse mammary tumour virus (MMTV), present in the genome of most mice. The transmission of the virus can be vertical, by genetic inheritance of proviral copies, or horizontal, by virus infection in the milk. As with

ALV, MMTV will not transform cells in culture and it does not carry an oncogene. It does, however, have a gene *orf* (open-reading frame) in addition to *gag*, *pol* and *env* (Kennedy *et al.*, 1982) but this is not derived from host cellular sequences and no *orf* protein product has yet been detected.

The mechanism of tumour induction by MMTV is also thought to involve insertional mutagenesis. Two groups have cloned mouse mammary tumour proviruses from tumours and analysed sequences flanking the virus. One group showed that, in the C3H strain of mice, 70% of tumours contained MMTV integrated at the same locus, called *int-1* (Nusse and Varmus, 1982). Another group showed that in the BR6 strain about 50% of tumours had virus integrated at another quite distinct locus, *int-2* (Peters *et al.*, 1983). In both cases the virus is found integrated on either side of *int* such that the transcriptional sense of the virus is away from the gene. Neither *int-1* nor *int-2* RNA has been found in normal mammary tissue but after MMTV integration transcription from one or other of the genes can be detected. The exact mechanism of viral activation of the locus is unknown. Activation results in only about ten copies of *int-1* RNA per cell but it can occur even when the virus is integrated 8 kb away. One possible explanation is that the virus contains a tissue-specific enhancer, presumably specific for mammary epithelial cells. It is known that the MMTV LTR contains an enhancer the activity of which is dependent on steroid hormone stimulation (Lee *et al.*, 1981) and in the lactating mammary tissue this enhancer may be activated. The consequences of this for the cell, i.e. possible production of orf and/or int-1–int-2 proteins, are only just being addressed.

int-1 and *int-2* genes are both present in the human genome. They have no homology with known viral oncogenes (Van Ooyen and Nusse, 1984) and because there is no way to assay for the effects of the genes it is still far from clear how they function. However, they clearly play a role in the initiation of MMTV-induced mouse mammary tumours and as such should be regarded as proto-oncogenes.

We have looked at two ways in which retroviruses can initiate tumour formation in animals: by carrying their own viral oncogene(s) or by insertional *cis* activation of a proto-oncogene. In the final section on viruses, we shall look at a third mechanism.

3   Human T-lymphotropic viruses

The human T-lymphotropic virus HTLVI is the only convincing

example so far of a retrovirus acting as an aetiological agent in a human malignancy. HTLVI is associated with a rare form of adult T-cell leukaemia and the first line of evidence that a virus might be involved came through epidemiological studies. The disease was found to be clustered in Japan and the West Indies and a virus was eventually isolated from a patient in the USA (Poiesz *et al.*, 1980). Two members (I and II) of the family have been identified in and isolated from tumour tissue and both are capable of transforming normal T lymphocytes in culture (Popovic *et al.*, 1983).

The genome of HTLVI has been completely sequenced (Seiki *et al.*, 1983). At the 3′ end is an open-reading frame which would code for a protein of 40 000 daltons (pX); this has been detected in infected cells using antibodies directed against synthetic peptides from the pX region (Slamon *et al.*, 1984b). However, the pX gene is not derived from cellular sequences and HTLVI therefore does not contain a viral oncogene analogous to the acutely transforming retroviruses. Unlike ALV and MMTV, there are no specific sites for integration of HTLVI (Seiki *et al.*, 1984). A clue to how the virus might initiate transformation has come from studies using the HTLVI LTR linked *in vitro* to an indicator gene, that encoding chloramphenicol acetyl transferase (CAT) (Sodroski *et al.*, 1984). CAT production from the viral LTR promoter could not be detected when the construct was introduced into a variety of cells but it could be detected in HTLVI-infected cells. It has been proposed that the HTLVI provirus produces a protein (perhaps pX) which activates transcription from its own promoter. It is possible that this HTLVI transcriptional factor can act in *trans* on specific cellular genes which, together with subsequent events, might result in tumour formation.

D   Non-viral activation of proto-oncogenes

The extensive knowledge gained about RNA tumour viruses and their mode of action has pointed those interested in human cancer firmly in the direction of proto-oncogenes. With the exception of HTLVI (and it may be that more viruses are yet to be discovered) human malignancies are thought to be caused by environmental agents such as diet and life style. These agents must still exert their effect at the genetic level and some of the 30 or so known proto-oncogenes might be involved. In the rest of section IV, we shall look at the massive research effort undertaken in the last five years to identify genes important in human cancer.

1   Changes in gene expression

The possibility that any one of the known cellular proto-oncogenes may be abnormally expressed in a tumour cell has been pursued by many groups (see, for example, Eva *et al.*, 1982, Westin *et al.*, 1982b, and Slamon *et al.*, 1984a). The results obtained are far from consistent and are often difficult to interpret, since many of these genes are expressed in normal cells. It was reported that high levels of *c-sis* RNA, which we now know codes for one of the chains of PDGF, are present in sarcomas and gliomas, whereas no expression was found in any other tumour or normal cell examined (Eva *et al.*, 1982). Other groups, however, have failed to confirm a high frequency of *sis* expression associated with fibrosarcomas (Slamon *et al.*, 1984a). Fibroblasts and glial cells are known to be responsive to this growth factor and it is possible that PDGF might be playing a role in tumorigenesis by an autocrine effect. Furthermore, it has been shown that if *c-sis* cDNA is linked to a strong promoter it is capable of morphologically transforming an established mouse fibroblast cell line, NIH3T3 (Clarke *et al.*, 1984; Gazit *et al.*, 1984), suggesting that *c-sis* can function as an oncogene. However, there is no evidence to suggest any genetic alteration at the *c-sis* locus in sarcomas and gliomas (Ratner *et al.*, 1985) and it is still not clear whether *c-sis* expression in these cell lines is a cause or an effect of transformation.

The following observations seem to be generally agreed: (a) *c-myc* and *c-ras* are expressed in almost all tumour and normal cells although there are some reports of increased transcription in some tumours; (b) *c-src*, *c-mos*, *c-rel*, *c-sis* and *c-yes* RNAs are detected in very few cell types; (c) *c-fes* and *c-myb* RNAs are often found in tumours but mainly in haematological malignancies, although the significance of this is not apparent as they are often expressed in normal haematopoietic cells. It is too early to assess the significance of these results but they do represent a framework for further analysis. If increased expression of a proto-oncogene could be linked to a genetic change, this would be a much stronger indication of an important event in the development of a tumour. We shall now look at some results obtained by analysing the proto-oncogenes themselves.

2   Gene amplification

One way to achieve high level expression of a proto-oncogene is by increasing its copy number. Gene amplification has been studied experimentally, notably for genes that confer a readily selectable phenotype. For example, an increase in the levels of the cytotoxic drug methotrexate, which is an inhibitor of dihydrofolate reductase (DHFR), results in resistant cell lines containing up to 100 copies of the DHFR gene (Schimke *et al.*, 1978). A stretch of DNA of approximately 1000 kb is the amplified unit and these large amplified sequences can often be detected cytologically, either as small independently replicating chromosomes called double minute chromosomes (DMCs) or as contiguous stretches, called heterogeneously staining regions (HSRs), integrated into a chromosome (Schimke, 1982). The HSRs need not be on the same chromosome as the original gene, suggesting that amplification might occur via DMCs which can then reintegrate to give HSRs.

HSRs and DMCs have long been known to occur in human tumour cell lines, suggesting that a gene involved in tumorigenesis might be amplified (Cowell, 1982). The first human cell line reported to contain an amplified oncogene was the promyelocytic leukaemia line HL60 (Collins and Groudine, 1982). About 20 copies of the *c-myc* gene are present in this line with concomitant high levels of *c-myc* RNA. However, this does not appear to be a common feature of this or any other type of leukaemia. Almost all examples so far reported are sporadic: amplified copies of *c-myb, c-abl, c-erbB* and *c-Ki-ras*, for instance, have been detected in a handful of tumour lines and in most cases the genes are located on DMCs or HSRs (or both) (e.g. Schwab *et al.*, 1984, and Libermann *et al.*, 1985). The only consistent pattern of gene amplification in tumours observed so far is of a novel gene with sequence homology to *c-myc* in human neuroblastomas (Schwab *et al.*, 1983). This gene has been called *N-myc* and the high frequency with which *N-myc* amplification is observed in these tumours suggests that it contributes to the tumour phenotype. It can therefore be tentatively classified as a cellular oncogene.

3   Gene rearrangement

Chromosomal rearrangements such as translocations, inversions and deletions are frequently observed cytologically in human tumours. Almost all the cellular proto-oncogenes have been chromosomally mapped (see Table 3) and much work is being done to see whether they are affected by these abnormalities.

In solid tumours, i.e. sarcomas and carcinomas, chromosomal changes are often heterogeneous and not well defined. This makes them difficult to analyse, although there are a few well-characterized cases.

Table 3 Human proto-oncogenes.

	Detection[a]	Chromosomal location in humans	Involvement in human cancer
c-src1	V	20	—
c-fps (fes)	V	15q25–q26	—
c-yes1	V	18q21.3	—
c-ros	V	—	—
c-abl	V	9q34	Translocated in chronic myeloid leukaemias
c-fgr	V	1	—
c-fms	V	5q34	—
c-erbB	V	7	Amplification in gliomas
c-raf (mil)	V	3p25	—
c-mos	V	8q22	—
c-sis	V	22q12–q13	High expression in sarcomas
c-Ha-ras1	V	11p15	Point mutations in many tumours
c-Ki-ras2	V	12p12–pter	Point mutations in many tumours
N-ras	T	1pcen–p21	Point mutations in many tumours
c-fos	V	14q21–q31	—
c-myc	V	8q24	Translocated in Burkitt's lines
c-myb	V	6q22–q24	Amplification
c-ski	V	1q12–qter	Amplification —

B-lym	T	1p32	Activated in B-cell lymphomas
c-rel	V	—	—
T-lym	T	—	Activated in T-cell lymphomas
c-met	T	7q21–q31	(Activated in a cell line)
c-neu (c-erbB2)	T	17q21	Amplification
c-ets1	V	11q23–q24	—
c-ets2	V	21q22.3	—
c-erbA	V	17	—
N-myc	c-myc homology	2p23–pter	Amplification in neuroblastomas
int-1	MMTV integration	12q14–qter	—
int-2	MMTV integration	—	—
bcl-1	B	11q13	Translocated in some B-cell leukaemias and/or lymphomas
bcl-2	B	18q21	Translocated in some B-cell leukaemias and/or lymphomas
tcl-1	B	14q32.3	Translocated in some T-cell leukaemias and/or lymphomas

[a] V, by homology to a viral oncogene; T, by the transfection assay; B, by localizing translocation breakpoints.

For example, a deletion on the short arm of chromosome 11 at band 13 (11p13) is associated with aniridia which predisposes patients to Wilm's tumour (a kidney tumour). This deletion is close to the *c-Ha-ras 1* proto-oncogene although as yet there is no evidence that this gene is directly affected. Chromosomal deletions not only may affect gene transcription and control but also could result in gene loss or inactivation. It has been suggested that some oncogenes may be recessive, such that in addition to the alteration of one copy of a gene the normal copy must be inactivated. Alternatively, it is possible that both copies of a gene must be inactivated. Some strong evidence for this proposal has come from the hereditary form of retinoblastoma where a deletion at 13q14 (band 14 on the long arm of chromosome 13) is sometimes observed as an inherited trait. In retinoblastoma cells obtained from these patients the normal chromosome 13 is missing, suggesting that both copies of a gene present at 13q14 must be removed (Benedict *et al.*, 1983).

In contrast with solid tumours, specific chromosomal translocations have been extensively documented in haematopoietic malignancies (Rowley, 1980). Human Burkitt's lymphomas which are often, although not always, associated with the Epstein–Barr virus, have one of three translocations: 8;14 (the end segments of the long arms of chromosomes 8 and 14 are exchanged), 8;2 or 8;22, with 8;14 accounting for about 90% of cases (Klein, 1983). It has now been shown that the *c-myc* gene is present on the fragment of chromosome 8 that is translocated and that the point of attachment to chromosome 14 is at the immunoglobulin heavy chain locus (Fig. 4). In the 8;2 and 8;22 cases, the κ and λ

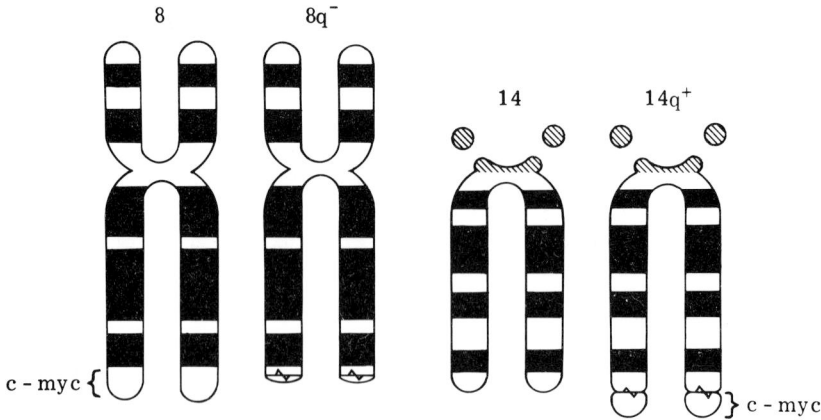

Figure 4 Schematic diagram showing an 8;14 chromosomal translocation. The banding pattern is observed by staining metaphase preparations of chromosomes.

immunoglobulin light chain loci are translocated from chromosomes 2 and 22 respectively to the *c-myc* gene on chromosome 8. Together with the fact that B cells are known to express immunoglobulin genes, this is compelling evidence that rearrangements at the *c-myc* locus are an integral part of the development of human B-cell tumours. An analogous situation occurs in mouse plasmacytomas, an equivalent murine disease.

By analogy with ALV insertional activation of the *c-myc* locus in chicken bursal lymphomas, it might be expected that the translocations in Burkitt's lymphoma would result in high level expression of the gene. However, the biochemical consequences of *c-myc* translocations have turned out to be extremely complicated and somewhat controversial. Nishikura *et al.* (1983) transferred a chromosome containing either a translocated or a normal *c-myc* gene into mouse plasmacytoma cells by making use of somatic cell hybrids. The hybrids containing the normal gene did not express *myc* mRNA whereas hybrids containing the translocated allele did. When cell hybrids were made with a mouse fibroblast line, however, neither allele gave significant *myc* transcription. They suggest that *c-myc* expression is increased about twentyfold by the translocation event but only in restricted cell types, i.e. B cells.

However, other groups have shown that the situation might not be so simple and that many Burkitt lines have very little (twofold) enhancement of *myc* expression. Both the immunoglobulin and *c-myc* loci are complicated (Fig. 5) and the breakpoints vary in different cell lines. Accordingly, the mechanistic interpretations that have been proposed differ, depending on where these breakpoints occur. The *myc* gene consists of three exons, although exon I is not translated into protein (Watt *et al.*, 1983). Within exon I, however, there is a 68 bp sequence which is almost exactly complementary to 70 bp in exon II. It has been proposed that in *c-myc* mRNA these two sequences can anneal to give a stem and loop structure which is only poorly translated into protein (Saito *et al.*, 1983). Some Burkitt lines (e.g. Manca, see Fig. 5) involve translocations where the breakpoint separates the *c-myc* promoter and exon I from exons II and III (Battey *et al.*, 1983). It is possible that in these cases the RNA produced by the translocated *c-myc* gene is translated much more efficiently into protein.

Increased transcription in the majority of lines in which the exon structure remains intact (e.g. BL22, see Fig. 5) is more difficult to explain. As shown in Fig. 5, the orientation of *c-myc* and the IgH locus in the 8;14 translocations is head to head (5′ to 5′), so that the immunoglobulin promoter is in the wrong orientation to affect *c-myc* expression. Furthermore, in at least one 8;2 translocation the immunoglobulin light

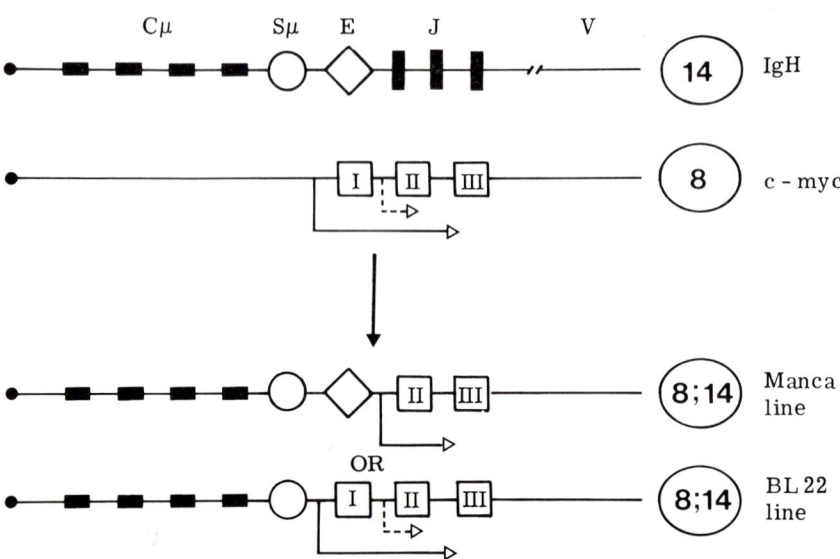

Figure 5 The IgH and *c-myc* loci before and after translocation. IgH is on chromosome 14 which is shown with the centromere (●), the constant-region exons (Cμ), the switch region (Sμ), an enhancer element (E), the joining-region exons (J) and the upstream variable region (V). Transcription of the IgH locus is from the V region into the C region. The *c-myc* gene is on chromosome 8 which is shown with the three exons (exon I, non-coding; exons II and III, coding), the normal mRNA start point (——) and the position of cryptic promoters (- - -). Two Burkitt lymphoma cell lines that have been analysed (Manca and BL22) are shown. Manca has the first exon of *c-myc* removed and the enhancer element of the Ig locus is thought to promote transcription from the cryptic promoters. The loss of exon I may also be significant (see text). In BL22 exon I remains but the enhancer is lost; it is not known how *c-myc* expression is affected in this line, although exon I does contain a few somatic mutations.

chain gene is positioned over 100 kb 3′ of the *c-myc* locus (Emanuel *et al.*, 1984). One plausible explanation proposed to explain some of these phenomena is that transcriptional enhancers within the immunoglobulin loci are responsible for either increased or, perhaps more importantly, uncontrolled transcription (Hayday *et al.*, 1984). A further possible effect is that of somatic mutation. The immunoglobulin loci are known to be susceptible to somatic mutations (Tongegawa, 1983), this being one of the mechanisms for the generation of antibody diversity (see Steinmetz, this volume). In a few Burkitt's lines (e.g. BL22) but by no means all, mutations in the *c-myc* gene, and in particular in exon I, have been found (Rabbitts *et al.*, 1983). The possible significance of this stems from the observation that the unrearranged *c-myc* gene in Burkitt's lines is not expressed, and Leder *et al.* (1983) and Rushdi *et al.* (1983) have proposed that *c-myc* transcription might be controlled by a *trans*-acting repressor which binds to exon I. Alterations in this exon as a consequence of translocation could prevent binding of the repressor, leading to uncontrolled expression; the untranslocated gene would have a normal exon I and would remain silent. It should be clear that much is still to be learned about the control and regulation of the *c-myc* gene itself. Such knowledge might help to explain how activation of the locus can be achieved by a seemingly endless variety of translocations.

About 90% of chronic myeloid leukaemias are associated with the Philadelphia chromosome which involves a reciprocal translocation between chromosomes 9 and 22. It has been shown that the *c-abl* gene is transferred from chromosome 9 to 22 and that the *c-sis* gene is transferred from chromosome 22 to 9 (Heisterkamp *et al.*, 1983). However, no expression has been detected from *c-sis* whereas the appearance of a novel *c-abl* transcript of 8 kb correlates extremely well with the presence of a Philadelphia chromosome (Gale and Canaani, 1984). The region of attachment of the *c-abl* gene to chromosome 22 has been cloned and analysed in detail (Heisterkamp *et al.*, 1985) and corresponds to a gene, *bcr* (breakpoint cluster region), which is normally expressed in myeloid cells. As a result of the translocation a hybrid RNA molecule is transcribed which contains some coding exons from *bcr* and all coding exons except the first from *c-abl*. The result of this 9;22 translocation in chronic myeloid leukaemias, therefore, is the production of an abl protein with a novel amino terminus derived from the *bcr* gene. This amounts to strong evidence that alterations at the *c-abl* locus are an important contribution to the development of chronic myeloid leukaemias.

Translocations involving chromosome 14 are found in other types of human leukaemia and lymphoma. Some types of B-cell lymphoma are associated with 11;14 or 14;18 translocations and in both cases the immunoglobulin heavy chain locus is involved (Tsujimoto *et al.*, 1985). The points of attachment to chromosomes 11 and 18 have been cloned and these loci (*bcl-1* (on chromosome 11) and *bcl-2* (on chromosome 18)) are regarded as being sites of putative oncogenes.

Some T-cell leukaemias and lymphomas are associated with either a 14;14 translocation or an inversion within 14. In these cases it is the α-chain gene of the T-cell receptor (located at 14q11) which appears to be involved (Croce *et al.*, 1985). Croce *et al.* have tentatively defined an oncogene *tcl-1* located at the tip of the long arm of chromosome 14. It is postulated that this gene is activated by translocation to the T-cell receptor α-chain locus and that this can occur by a 14;14 translocation or by an inversion within the long arm of 14.

E   Biological assays to detect cellular oncogenes

1   *NIH3T3 transfection*

So far we have concentrated on known proto-oncogenes in human tumour cells. Several years ago two groups reported the detection of cellular oncogenes using a biological assay (shown schematically in Fig. 6) without prior knowledge of what they might be (Shih *et al.*, 1979; Cooper *et al.*, 1980). They made use of an established mouse fibroblast cell line, NIH3T3, which lacks all the properties normally associated with transformed fibroblasts (see Table 1) except that of immortality. These cells can take up exogenously added DNA in the form of a calcium phosphate precipitate, a technique first described by Graham and van der Eb in 1973, and the DNA is stably integrated into the mouse chromosomes. If DNA from tumour cells is used in the assay, then a proportion of the NIH3T3 cells become morphologically transformed and these can be observed either as a focus of cells on a confluent monolayer background or as colonies in soft agar (see Fig. 6). Furthermore, when these transformed cells are picked and grown, they give rise to tumours when injected into nude mice. It appears that genetic information, i.e. a cellular oncogene, is being transferred from a tumour cell into NIH3T3 causing transformation.

In 1981 the same two groups reported that DNA isolated from a human tumour cell line, EJ (a bladder carcinoma, also called T24), would transform NIH3T3 cells, whereas DNA from normal human cells would not; they had detected, using a biological assay, a human cellular oncogene (Krontiris and Cooper, 1981; Shih *et al.*, 1981). Many groups were involved in identifying and cloning this gene. One cloning strategy adopted was to make use of the Alu repeat sequence present in human DNA (Shih and Weinberg, 1982). There are about 300 000 copies of this repeated unit distributed randomly throughout most of the human genome and it has been estimated that during the transfection procedure each NIH3T3 cell takes up about 0.1% of the human genome. Primary foci of transformed NIH3T3 cells are therefore expected to contain human DNA equivalent to about 1000 genes, including any Alu repeat sequences associated with them. If DNA is isolated from such a primary focus and used in a second round of transfection, then the secondary foci produced should contain 0.0001% of the human genome,

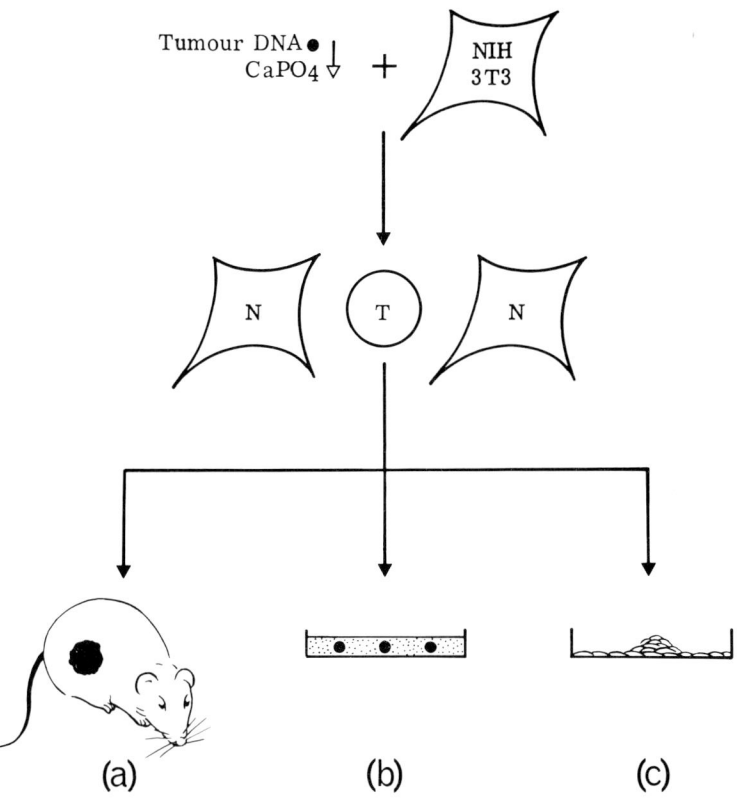

Figure 6   The NIH3T3 transfection assay. A calcium phosphate precipitate of high molecular weight tumour DNA is added to NIH3T3 cells in tissue culture. After about 2 weeks some transformed cells (T) can be detected within the population of normal cells (N). These are observed by tumour formation after injection into nude mice (a), colony formation in soft agar (b) or focus formation on a confluent monolayer (c).

i.e. one gene. In practice, a handful of sequences is found and the transforming gene must be one of them. It is clear that, if this gene is associated with a repetitive element, it too will be present in the secondary focus and this can be detected on Southern blots using the cloned Alu sequence as a probe (Fig. 7). In the case of the EJ transforming gene, only one fragment was conserved in *all* secondary foci examined and it was concluded that this either contained or was closely linked to the oncogene. The fragment was cloned and it did indeed contain a biologically active oncogene; it would transform

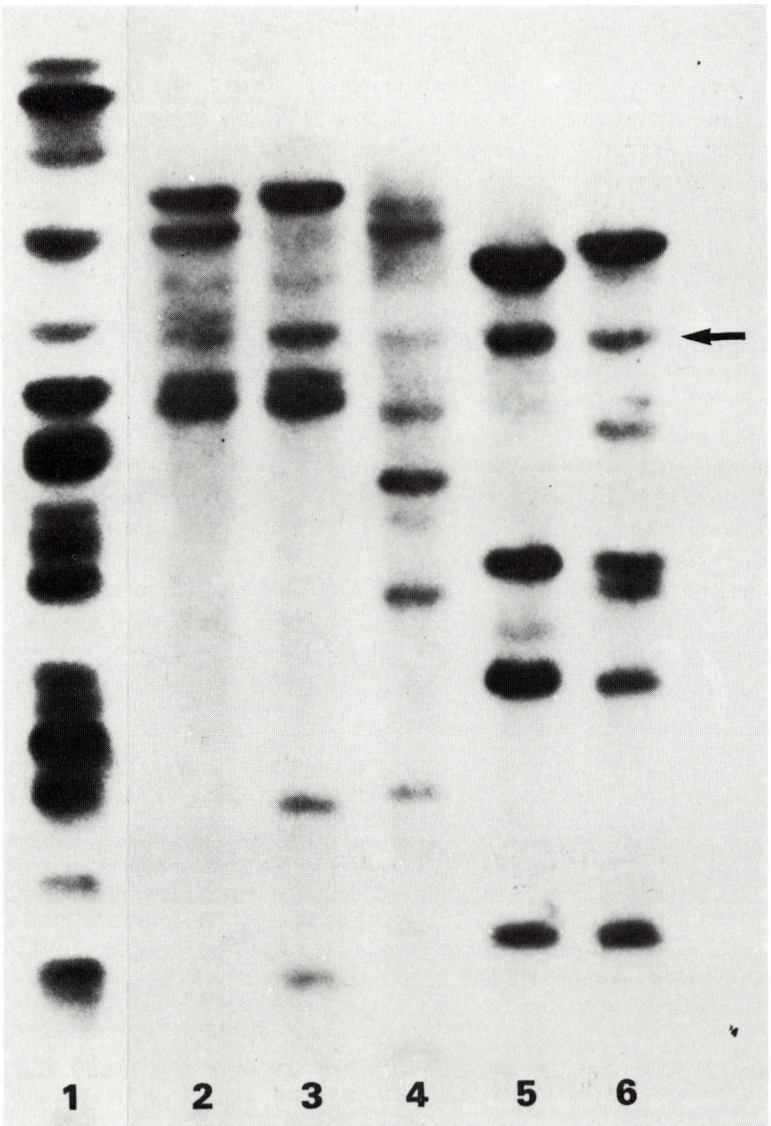

Figure 7 An example of a Southern blot showing *Eco*RI-digested DNA obtained from a primary focus (track 1) and from five independent secondary foci (tracks 2–6) hybridized with a human Alu repetitive-sequence probe. The primary focus contains many bands but these are reduced to less than ten in the secondaries. Only one fragment (marked by an arrow) is conserved in all secondary foci. It can be concluded therefore that this fragment is part of, or very closely linked to, the transforming gene.

NIH3T3 cells several thousand times more efficiently than total EJ DNA.

It soon became apparent that this gene was related to a known viral oncogene, the viral *ras* gene; it was the human *c-Ha-ras1* proto-oncogene (Santos *et al.*, 1982). The *c-Ha-ras1* gene had already been cloned from normal human DNA (Chang *et al.*, 1982b) and it is not capable of transforming NIH3T3 cells unless it is first attached to a strong promoter (Chang *et al.*, 1982a). This shows that inappropriate high level expression of the normal gene can lead to transformation and with this in mind the EJ cell line was analysed for either amplified or rearranged *c-Ha-ras1* or for increased gene expression. None of these was observed. To locate exactly the position of the alteration it was decided to make *in vitro* recombinants, replacing restriction fragments present in the normal non-transforming gene with corresponding fragments from the EJ gene (Fig. 8). These chimeric molecules were then tested in the transfection assay and eventually a small 350 bp fragment from the EJ gene was found to activate the normal gene (Reddy *et al.*, 1982; Tabin *et al.*, 1982). This fragment was then sequenced from the normal and the transforming versions of the gene and only one base change was found. This single alteration, out of the 6600 bp constituting the *c-Ha-ras1* gene, is sufficient to convert the proto oncogene into a cellular oncogene. The point mutation resulted in a

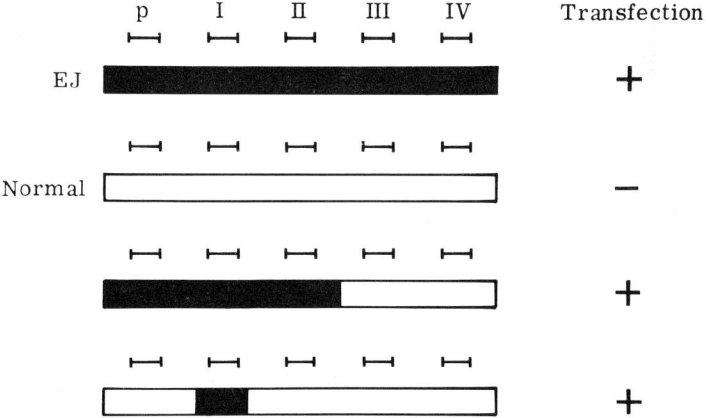

Figure 8   Use of chimeric molecules to locate the activating lesion in the EJ version of *c-Ha-ras1*. The recombinant clones were tested for transforming activity on NIH3T3 cells (+, active; −, inactive). The approximate positions of the promoter region (p) and the four protein coding exons (I–IV) are also shown.

change of the twelfth amino acid of the 21 000 dalton ras protein (p21) from a glycine to a valine residue.

Since the EJ cell line was first used in the transfection assay, well over 100 different human tumours have been tested. Around 20% score positive and in almost all cases the oncogene belongs to the *ras* gene family. The *c-Ha-ras1* gene is only rarely detected. An activated *c-Ki-ras2* gene is detected much more frequently, particularly in lung and colon carcinomas but also in leukaemias and other tumours. At that time four members of the *ras* gene family were known (two of which, *c-Ha-ras2* and *c-Ki-ras1*, are pseudogenes) (for a review of the *ras* gene family see Hall, 1985); however, characterization of a new oncogene detected in a variety of tumour lines showed that it too was related to *ras* (Hall *et al.*, 1983). It has been called *N-ras* (Shimizu *et al.*, 1983). Activated *N-ras* is detected frequently in leukaemias but is also found in carcinomas and sarcomas. Interestingly, an active *N-ras* gene has been detected in the HL60 cell line which also contains amplified *c-myc* sequences (Murray *et al.*, 1983) and altered *ras* genes have been detected in some Burkitt cell lines together with a translocated *c-myc* gene. It is possible that these represent two of the genetic alterations in the development of these tumours.

For all three genes the mechanism of activation always involves a point mutation at amino acids 12, 13 or 61 and we shall look at the biochemical basis of this activation in section V. Normal cells, taken from a patient with a tumour containing a mutant *ras* gene, do not contain these alterations (Gambke *et al.*, 1984); the point mutation is therefore a somatic event and environmental carcinogens are the prime suspects.

Other oncogenes have been detected using the NIH3T3 assay. Lane *et al.* (1982) have reported a variety of human and mouse leukaemia-specific oncogenes but to date only two of these, *B-lym* (from intermediate B-cell leukaemias) and *T-lym* (from intermediate T-cell leukaemias), have been cloned (Diamond *et al.*, 1983; Lane *et al.*, 1984). Very little is known about them except that they are not related to any of the known viral oncogenes. An activated *B-lym* gene is also detected in DNA from chicken bursal lymphomas which, as we have already seen, are induced when ALV integrates at the *c-myc* locus. However, in the transfection assay it is a *B-lym* gene which is detected, not *c-myc* (Goubin *et al.*, 1983), and the *B-lym* locus is presumably the site for a second event in the development of these tumours. An oncogene *c-neu* has been detected in experimentally derived rat neuroblastomas (Schecter *et al.*, 1984) and, although it has homology with *v-erbB*, *c-neu* is distinct from the *c-erbB* gene. An activated *c-met* gene has been found in an experimentally transformed human osteosarcoma cell line (C. S. Cooper *et al.*, 1984) but nothing is yet known about the properties of this gene.

Table 3 shows a current list of almost 30 proto-oncogenes present in the human genome. It can be seen that there is evidence for involvement in human cancer for about a dozen of them. The NIH3T3 transfection assay has so far not led to the discovery of an extensive set of new cellular oncogenes, as once hoped. However, because it is a biologically based assay, it has proved beyond reasonable doubt that cellular oncogenes do exist in human tumours and this has opened the door to a wide range of experiments on the molecular biology of human cancer.

The NIH3T3 assay does have serious limitations, the most intriguing being that still only 20% of human tumours contain a detectable oncogene. Two explanations for this immediately spring to mind. First, it is possible that certain alterations in the *ras* genes might go undetected and indeed the recently discovered position 13 changes are poor at eliciting transformed foci in the NIH3T3 transfection assay (Bos *et al.*, 1985). These so-called weakly transforming alleles were detected by injecting the transfected NIH3T3 cells into nude mice and looking for the appearance of tumours. It may be that more sensitive biological assays such as tumour formation, or biochemical techniques such as oligonucleotide probes (Bos *et al.*, 1984) and monoclonal antibodies capable of distinguishing small changes in DNA or in protein, will uncover many more altered *ras* genes in human tumours. Second, it is possible that in 80% of human tumours oncogenes other than *ras* are involved and that they are not detectable in NIH3T3. Attempts have been made to use other recipient cells but with little success. The major problem is the inefficiency with which other cells take up DNA and experiments designed to increase DNA uptake are being pursued.

2   *Transformation of other cells*

We must not forget that human cancer is a multistep process and that even in cells with an activated *ras* oncogene other genes are likely to be involved. The *ras* mutation is only one step in tumour formation and, as we have seen, in some Burkitt's tumours *c-myc* and *c-ras* might both be involved. It was not too surprising, therefore, when it was shown that the cloned EJ oncogene could not transform primary cell cultures. However, a breakthrough was reported in 1983 when two groups showed that non-established cells (rat embryo fibroblasts or baby rat kidney cells) could be transformed using combinations of cloned oncogenes (Land *et al.*, 1983; Ruley, 1983). The activated *ras* genes could be combined with any one of the following: *myc*, polyoma large T-antigen gene, adenovirus E1a gene and, more recently, the cellular p53 gene (Parada *et al.*, 1984). These combinations give rise to transformed primary cells in an apparently two-step process. What have these genes in common? First, they are all genes coding for nuclear proteins, unlike the ras protein which is cytoplasmic. Second, E1a and large T-antigen have previously been shown to have an immortalizing effect on cells (Houweling *et al.*, 1980; Rassoulzadegan *et al.*, 1983) and it was concluded that any one of these nuclear proteins could immortalize cells, the ras proteins producing the full transformed phenotype.

These intriguing experiments have caused a great deal of excitement but the exact interpretation of the results is still far from clear. For instance, there is no evidence that either *v-myc* or *c-myc* can immortalize cells; the MC29 virus does not immortalize cells and Epstein–Barr virus can immortalize B cells to form lymphoblastoid cell lines without the *c-myc* translocation (Pope *et al.*, 1973), suggesting that alterations to *c-myc* in Burkitt's lymphomas are not involved in immortalization. Furthermore, one group has even reported that very high levels of mutant ras protein can both immortalize and transform early passage Chinese hamster lung cells (Spandidos and Wilkie, 1984) and others have shown that *myc* can morphologically transform the already established NIH3T3 cell line (Vennstrom *et al.*, 1984). Some of the controversy might be accounted for by the fact that different cells have been used with a variety of plasmid constructs producing different levels of oncogene products; the exact meaning of the two-step transformation of primary cells remains unclear. Most groups agree that normal levels of *ras* or of *myc* cannot by themselves immortalize primary cells; however, in combination they clearly produce immortal transformed cell lines. It seems likely that some other secondary events are occurring in the cotransfection experiments and the search for a better understanding of the basis of immortalization is becoming more and more important. The possibility that transfection of primary cells with total human tumour DNA in the presence of a cloned *ras* gene might lead to the identification of a new set of human tumour genes has not escaped notice. However, primary cells are relatively inefficient at taking up DNA and there are no reports yet of any success.

Several groups have developed *in vivo* animal models where known carcinogens are administered to mice or rats, resulting in a reproducible high level occurrence of tumours. One group has shown that if the carcinogen dimethylbenzanthracene is applied to mouse skin it will give rise to skin tumours and in all cases the mouse *c-Ha-ras1* gene is activated (Balmain *et al.*, 1984). Similarly, if nitrosomethylurea is administered to rats breast carcinomas develop at a high frequency and these all have activated *c-Ha-ras1* genes (Zarbl *et al.*, 1985). Guerrero *et al.* (1984) have treated mice with either X-ray irradiation or chemical carcinogens to generate thymic tumours. Again, *ras* gene activation occurs but interestingly the irradiated mice have an altered *c-Ki-ras2* gene, whereas the carcinogen-treated mice have an altered *N-ras* gene. These experiments suggest that *ras* activation might be an early event in carcinogenesis and a direct consequence of the presence of the carcinogen. In contrast, injection of nitrosoethylurea into pregnant rats results in a high incidence of neuroblastomas in their offspring and in this case there is no evidence for *ras* alteration (Schechter *et al.*, 1984). Instead a new oncogene, *neu*, was identified using the NIH3T3 transfection assay.

In addition to immortalization, another important aspect of human tumours is their capacity to metastasize. Some preliminary results have shown that if activated *ras* genes are introduced into non-metastasizing mouse tumour lines these acquire a much higher metastatic capability (Vousden and Marshall, 1984; C.J. Marshall and K.H. Vousden, personal communication). It is thought that metastasis is a late event in tumour progression and it is possible that some of the observed *ras* alterations have occurred after the actual transformation process.

ras is most definitely implicated in many human tumours. In a variety of experimental systems it has been shown to be capable of doing everything from immortalization, through morphological transformation, to contributing to the metastatic potential of already-transformed cells. Exactly when the *ras* mutation occurs in the development of human cancer and what its *in vivo* contribution is to the tumour phenotype is not known. It is hoped that the animal model systems and the *in vitro* experiments using both established and non-established cells will help to answer these questions.

V Biochemistry of oncogene proteins

A Cytoplasmic kinases

In 1978 Collett and Erikson succeeded in isolating the 60 000 dalton protein product of the *v-src* gene, pp60$^{v\text{-}src}$. They showed that it had an enzymatic activity, namely that, if the protein was immunoprecipitated in the presence of ATP, pp60$^{v\text{-}src}$ would phosphorylate the anti-src antibody. It was assumed at the time that this indicated that the protein was a serine- or threonine-specific protein kinase. However, it soon became apparent that it was phosphorylating tyrosine residues and this was the first protein tyrosine kinase identified (Hunter and Sefton, 1980).

We now know that pp60^{src} is one member of a family of oncogene products with tyrosine-specific protein kinase activity, which includes src, yes, abl, fps, fgr and ros (Fig. 9A). These proteins are all located at the inner surface of the cytoplasmic membrane (e.g. Courtneidge *et al.*, 1980) and a comparison of their amino acid sequences has shown that they are related to each other (Hunter, 1984). There is a stretch of about 250 amino acids in pp60^{src} which is the region responsible for the kinase activity and a corresponding domain is found in the other tyrosine kinases, with a high degree of amino acid conservation between them. This kinase domain is also found in the cytoplasmic cyclic-AMP-dependent serine protein kinase and in mos and raf, also serine-specific kinases located in the cytosol (Moelling *et al.*, 1984), indicating a distant evolutionary relationship between all these protein kinases. A similar sequence domain has been found in the membrane-bound products erbB, fms and neu and these all have a tyrosine kinase activity.

Clearly tyrosine kinases are a common feature of oncogene products and it was originally thought that the activity might be peculiar to oncogenes. This was soon disproved when a protein derived from the cellular *src* gene was isolated from normal cells (Collett *et al.*, 1979) and shown to have tyrosine-specific kinase activity. Since then other membrane-bound cellular proteins with similar activity have been identified and these have given a clue as to what the oncogene kinases might be doing. It has been demonstrated that the membrane receptors for the mitogens PDGF and IGFs each have a tyrosine-specific kinase activity (Cohen *et al.*, 1980; Ek *et al.*, 1982; Kasuga *et al.*, 1983) and it seems that tyrosine phosphorylation is a normal early event in the transduction of mitogenic signals through the membrane (Cooper *et al.*, Ek *et al.*, 1982). Although pp60^{c-src} is unlikely to be a receptor, since it is located on the inner surface of the membrane, it probably does play some role in this early signalling process. The presence of a tyrosine kinase encoded by a viral oncogene might therefore result in a continuous or uncontrolled mitogenic signal for cell division.

This still leaves unanswered the question of how these signals are read by the cell and many groups have attempted to find the cellular targets for phosphorylation by pp60^{src} and by the growth factor receptors. When RSV-transformed cells are examined many proteins show increased levels of phosphotyrosine, e.g. the cytoskeletal protein vinculin (Hynes, 1982), some glycolytic enzymes (Cooper *et al.*, 1983) and several proteins identified only by their molecular weight. However,

A *Growth factors* (1)
sis (PDGF)

Growth factor receptor-like proteins (2)
erbB (EGF receptor), fms, neu

G-like regulatory proteins (3)
Ha-ras, Ki-ras, N-ras

Tyrosine-specific protein kinases (4)
src, fps, yes, ros, abl, fgr

Serine/Threonine-
specific protein kinases (5)
mos, raf (mil)

Nuclear proteins (6)
myc, fos, myb, ski, B-lym

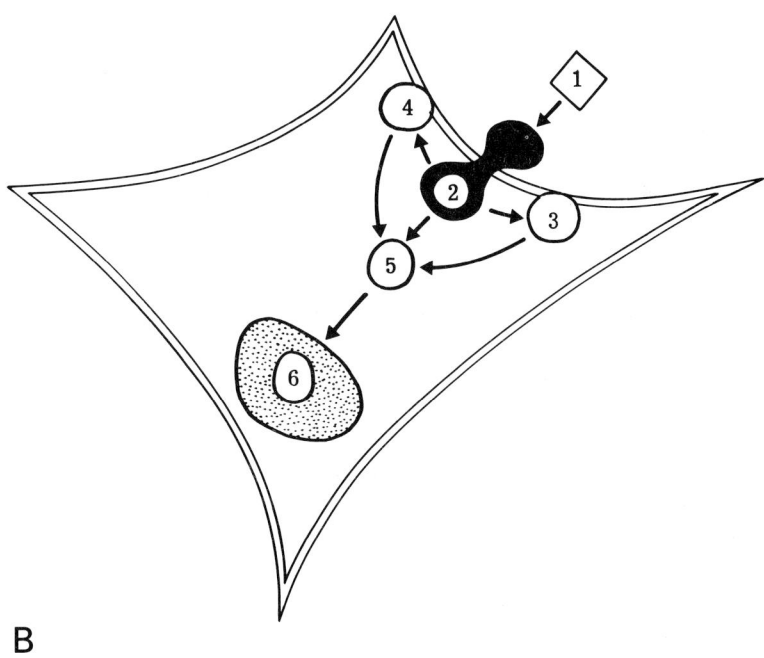

B

Figure 9 A, Activities of oncogene products. The available information has been used to place the oncogene products in groups. The protein kinases, for example, have similar enzymatic activities and they are sequence related. In contrast, no known function has been attached to the nuclear proteins, their location in the cell being the only thing they, so far, have in common. B, A highly schematic representation of how extracellular mitogenic signals are transmitted to the cytoplasm and nucleus. Examples of oncogene products are now known which could possibly fit into the different categories shown, and these are indicated in A and in the text.

there is no clear indication that any of these proteins are the immediate targets of the pp60^{src} kinase activity. Phosphorylation of a 42 000 dalton protein has been shown to be a common response in normal cells exposed to a variety of mitogens (J.A. Cooper et $al.$, 1984) and in transformed cells. A 36 000 dalton protein is also phosphorylated in RSV-transformed cells and this was thought to be a possible substrate for pp60$^{v\text{-}src}$. Recent work with plasmids containing v-src and a regulatable promoter, however, has shown that levels of pp60$^{v\text{-}src}$ just sufficient to transform cells do not result in phosphorylation of this protein (Jakobovits et $al.$, 1984). It would therefore seem that this step is not of primary importance in transformation. This type of experiment, using regulated levels of kinase activity through appropriate gene constructs, is likely to be of value in analysing which of the many changes seen in phosphotyrosine levels are secondary events and which are the initial targets. It should be said that secondary events are likely to be the cause of many of the characteristics associated with transformed cells; alterations in cytoskeletal proteins, for example, might be responsible for some of the changes in morphological and/or adhesion properties observed after transformation.

One effect of pp60$^{v\text{-}src}$ which is now thought to be of importance is that it leads to increased protein phosphorylation on serine residues (Decker, 1981). In particular, phosphorylation of the S6 ribosomal protein on a serine residue is thought to be a critical event in the mitogenic stimulation of normal quiescent cells. Clearly phosphorylation of S6 after RSV transformation would have to occur via at least one intermediate, namely a serine kinase which itself might be activated directly or indirectly by the pp60$^{v\text{-}src}$ tyrosine kinase (Blenis et $al.$, 1984). The cascade of events that transmits signals from the membrane through the cytoplasm to the nucleus is not known in any detail.

Two distinct biochemical pathways, however, have been recognized for several years as important in the mitogenic stimulation of cells and a possible interaction with both has been shown for pp60^{src}. Both pathways involve the generation of second messengers, i.e. molecules produced in response to an early event at the membrane which can then trigger a series of secondary events within the cell. The first involves the generation of cyclic AMP by membrane-bound adenylate cyclase. In response to agonists, some receptors, e.g. those for the β adrenergics and prostaglandins, stimulate adenylate cyclase leading to increased levels of intracellular cyclic AMP. This in turn can lead to activation of the cytoplasmic cyclic AMP-dependent serine-specific protein kinase which ultimately leads to activation of other serine-specific kinases, in particular protein kinase C (Flockhart and Corbin, 1982). It is protein kinase C which is thought to play a central role in the various cellular responses to mitogenic stimulation, including the phosphorylation of S6 (Nishizuka, 1984a). One group has shown that tyrosine phosphorylation by pp60$^{v\text{-}src}$ of the cyclic AMP-dependent protein kinase takes place in transformed cells, affecting its activity and regulation (Graziani *et al.*, 1984). It is therefore possible that the pp60$^{v\text{-}src}$ interacts with the pathway that regulates cell proliferation through cyclic AMP and protein kinase C.

The second pathway also leads ultimately to activation of protein kinase C. It is known that for full activity this protein requires two cofactors, Ca^{2+} and diacylglycerol (see Nishizuka, 1984b, for a review), both of which can be generated in response to a variety of extracellular signals, e.g. acetylcholine or the mitogen PDGF. The result of the interaction of these molecules with their receptors is a breakdown of inositol phospholipids located in the membrane to yield diacylglycerol, which activates protein kinase C, and inositoltriphosphate, which can affect Ca^{2+} levels within the cell. A link between transformation and this pathway came from the results of Sugimoto *et al.* (1984). They showed that pp60$^{v\text{-}src}$ could phosphorylate inositol phospholipids *in vitro* and that in RSV-transformed cells there is a build-up of intermediates in the inositol lipid breakdown pathway. They postulated that the primary target of pp60^{src} might be lipid and not protein. The same reactivity towards lipid has been shown for the ros product (Macara *et al.*, 1984).

There is still much to be learned about the biochemical action of pp60^{src} and the rest of the tyrosine kinase family. Phosphorylation of tyrosine seems to be a general phenomenon for initiating cell division and inappropriate tyrosine kinase activity could obviously explain the loss of growth control associated with transformed cells. The phosphorylation of inositol lipids by at least two of the tyrosine kinases (src and ros) is intriguing but its significance in transformed cells remains to be demonstrated. A schematic version of the effect of mitogens on cells is shown in Fig. 9B. A mitogen (1) interacts with its receptor (2), which in turn activates membrane-bound proteins (4). These could then transmit their signals to the cytosol via cytoplasmic serine kinases (5) or by phosphorylation of inositol lipids within the membrane. Inappropriate activation of type (4) proteins (c-src, c-yes, c-abl, c-ros, c-fps or even possibly adenylate cyclase) or type (5) proteins (c-mos, c-raf or possibly protein kinase C and other intracellular kinases) would obviously upset this delicate control system. However, this simplified model should not be taken as evidence that we understand how any of these products really work or how they are interconnected.

B   ras proteins

The 21 000 dalton (p21) products of the three human cellular *ras* genes (*Ha-, Ki-* and *N-ras*) are very closely related in sequence. In the first 150 amino acids there are a maximum of 14 amino acid differences between the three proteins, although in the C-terminal 39 amino acids they are about 50% divergent. Even more striking, the viral Harvey p21 protein and human *c-Ha-ras1* p21 differ in only three amino acids (see Hall, 1984, for a review). Recently, two *ras* genes have been identified in yeast (Defeo-Jones *et al.*, 1983; Powers *et al.*, 1984), and although the protein products are larger than the human equivalents (RAS1, 309 amino acids; RAS2, 322 amino acids), they are 90% homologous to amino acid residues 1–80 of the human p21s and 50% homologous to residues 81–160. The ras proteins therefore have been highly conserved throughout evolution and presumably they play a critical role in the regulation of cell growth.

The ras p21s are located at the inner surface of the plasma membrane and although the viral proteins are phosphorylated at amino acid residue 59 (a threonine) the human p21s do not have threonine at 59 and are not phosphorylated. As we saw earlier, the *ras* genes are activated by a point mutation in many human tumours and the mode of action of normal and transforming p21 is therefore of great interest. Both transforming (viral or activated cellular) and non-transforming (normal cellular) p21s bind GTP and GDP (but not ATP) equally and they both have a GTPase activity (McGrath *et al.*, 1984). The transforming proteins, however, hydrolyse GTP about ten times more slowly than the normal proteins do. What is the significance of this in terms of the role of the protein in cells? A clue to what p21 might be doing has come from understanding the role of other GTP-binding proteins in the cell.

Two classes of GTP-binding proteins (called G proteins) are involved in the regulation of the adenylate cyclase system already mentioned (see Gilman, 1984, for a review) and these are members of a larger family of GTP-binding proteins including transducin, a protein found in the retinal rod which regulates the levels of cyclic GMP. The way in which the proteins work is reasonably well understood at least in the adenylate cyclase system. Activation of certain receptors can stimulate one of the G proteins to bind GTP and it then interacts with and activates the adenylate cyclase catalytic subunit. This leads to an increase in the intracellular levels of the second messenger, cyclic AMP, which as we saw earlier can lead to the activation of intracellular kinases including protein kinase C. The G protein hydrolyses the GTP, leading to deactivation of adenylate cyclase in the absence of any further signal.

The grouping of the ras proteins in the G protein family is suggested by their biochemical activities and by some limited amino acid sequence analysis (Hurley *et al.*, 1984). It is therefore possible that the normal ras proteins also interact with a receptor and, in response to an external signal, bind GTP and interact with an as yet unknown molecule to generate a second messenger. Adenylate cyclase is unlikely to be directly involved as the G proteins associated with it are well characterized and have different molecular weights from ras p21. The reaction is controlled by hydrolysis of GTP and, since the transforming versions of p21 have reduced GTPase activity, this might result in abnormally high levels of the second messenger. Alternatively, they could generate this intracellular signal under conditions which are not mitogenic for normal cells, e.g. low serum, and this may be the basis for an uncontrolled cellular stimulus for growth. What the receptors might be and what the targets are is not known, and it is still not clear whether the three *ras* gene products behave differently. However, in the schematic diagram of Fig. 9B, we can suggest that the ras proteins correspond to a type (3) protein. If the p21s do interact with one of the tyrosine kinase growth factor receptors then this might regulate activation of other membrane kinases, leading to an increase in intracellular protein phosphorylation. Alternatively, some of the proteins involved in the inositol lipid breakdown pathway are GTP-binding proteins and it is possible that ras may be one of these.

The delicate genetic manipulations which are possible with yeast have allowed a more direct analysis of ras protein function in that organism than is currently possible in mammalian cells. Deletions introduced into the two yeast *ras* genes (Kataoka *et al.*, 1984) have shown that at least one *ras* product is essential for yeast spores to germinate, suggesting a crucial role in cell division. Furthermore, if one of the normal yeast *ras* genes is mutated in an analogous fashion to the mutants found in human tumour cells, the yeast strain fails to form spores in response to nutrient deprivation. These strains contain much higher levels of activated adenylate cyclase although it is not clear whether this is because of a direct interaction with ras or whether it is a secondary effect (Toda *et al.*, 1985).

Analysis of the role of ras in mammalian cells is going to be more complicated. Recently it has been reported that microinjection of anti-p21 antibodies into quiescent cells prevents them from entering the cell cycle even after serum stimulation (Mulcahy *et al.*, 1985). This agrees with the yeast analysis indicating that ras protein is essential for cell division. However, in order to progress in our understanding of ras protein function it will be essential to identify which proteins they interact with in the cell. Attempts to coprecipitate any associated proteins using anti-ras antibodies have been unsuccessful, implying either that any associations are weak or that they are dependent on intact membrane organization.

C   Nuclear proteins

The products of five oncogenes, *myc, myb, fos, ski* and *B-lym*, are known to be located in the nucleus (see, for example, Abrams *et al.*, 1982) but no biochemical activity has been attached to any of them. The expression of at least two of these products, myc and fos, appears to be dependent on the proliferative state of the cell. Quiescent 3T3 cells, for example, have undetectable levels of *fos* mRNA but within 30 min of stimulation by PDGF the levels are dramatically increased (about one hundredfold). However, this is only transient and after about 2 h they disappear (Muller *et al.*, 1984). Thus the interaction of PDGF with its receptor not only activates intracellular phosphorylation events and the breakdown of inositol lipids but also leads to the generation of a nuclear signal to switch on *fos* expression. Since phosphorylation of intracellular proteins occurs within a few minutes of mitogenic stimulation (Rodriguez-Pena and Rozengurt, 1985), it is likely that *fos* expression is a direct result of some of these events.

c-*myc* is expressed at very low levels in quiescent cells but this is increased up to twentyfold about 1 h after stimulation with PDGF (Kelly *et al.*, 1983; Muller *et al.*, 1984). The levels of expression then return to the basal level present in growing cells (about threefold to sixfold higher than in quiescent cells) and these are maintained throughout the cell cycle. Armelin *et al.* (1984) have taken this observation a step further by introducing into cells a *myc* gene linked to an inducible promoter. They showed that, when the *myc* gene is constitutively expressed, the cells no longer require PDGF for growth, confirming the hypothesis that *myc* induction is an important event in the stimulation of cell growth by PDGF.

In contrast with *myc*, which appears to be expressed in all cells, *myb* expression has only been found in haematopoietic cells. *B-lym* is expressed in most cells and it has some sequence homology to transferrin (Goubin *et al.*, 1983).

As to the role of the *fos* and *myc* gene products, very little is known. The myc protein has been shown to have DNA binding activity but no sequence specificity has been demonstrated and it is difficult to reconcile this with direct activation of specific genes. However, it has been proposed that myc could have a role in maintaining the continuous growth of cells. Expression is very low in quiescent cells and higher in growing cells (Campisi *et al.*, 1984) and, furthermore, when the promyelocytic leukaemia line HL60 is caused to differentiate terminally or when lymphoid cells are growth arrested by interferon, *myc* is again switched off (Westin *et al.*, 1982a; Jonak and Knight, 1984). It appears that myc is somehow involved in the transition of cells from quiescence (G0 phase) into the cell cycle (G1, S and M phases). fos expression appears to be an earlier and transient event; whether this protein activates specific genes (including *myc*) remains to be seen. Inappropriate expression (perhaps by genetic alterations of nuclear proteins (6) in Fig. 9) could keep the cell cycling even under conditions which would normally be sufficient to switch off cell growth and division.

D   Other oncogene products

As mentioned earlier, it has been shown that *v-erbB* is a truncated version of the EGF receptor gene (Downward *et al.*, 1984) and the *neu* gene detected by the transfection assay has homology with *erbB* and encodes a receptor-like molecule (Schechter *et al.*, 1984). Recently it has been shown that *v-fms* is derived from the cellular gene encoding the receptor for colony stimulating factor 1, a haematopoietic growth factor (Sherr *et al.*, 1985). In the case of *v-erbB* it is the extracellular EGF-binding domain which is lacking and an obvious possibility is that this truncated receptor is in an activated configuration even in the absence of EGF stimulation. One might predict the possibility that other known growth factor receptors such as those for PDGF or IGFs could be altered or inappropriately expressed in such a way as to yield oncogenic proteins, but so far no examples of this have been reported. The alteration of receptor molecules (Fig. 9, protein type (2)) completes the link between nucleus and membrane.

Finally, changes in the mitogens themselves (Fig. 9, protein (1)) or in their expression also appear to be possibilities. The *v-sis* sequences of the simian sarcoma virus correspond to the B chain of PDGF. PDGF is normally produced in a very restricted number of cell types, mainly bone marrow megakaryocytes, and receptors for the molecule have been found on mesenchymal and glial cells (for a review see Stiles, 1983). In the case of viral transformation of fibroblasts, the *v-sis* sequences are fused to the *env* sequences of the virus and this allows export of the PDGF-like molecule to the membrane. Presumably by autocrine stimulation through the PDGF receptor, inappropriate high level expression causes transformation of the cells. Abnormal expression of any mitogenic factor might be a possible candidate for a role in oncogenesis, provided that the cells expressing the mitogen carry the appropriate receptors. We saw in section I that many tumour cells release TGFs and one class of these, TGFα, is closely related in sequence to EGF and interacts with the EGF receptor (Derynck *et al.*, 1984). There is some evidence that the normal roles of TGF molecules are as mitogens necessary for early embryonic development (Sporn and Todaro, 1980; Twardzik *et al.*, 1982), and ectopic (inappropriate) expression in adult cells could be a step in transformation.

VI Summary

The study of oncogenes has come a long way since 1970. Tables 2 and 3 show that about 20 viral oncogenes and 30 cellular oncogenes have been identified to date and it has been suggested that there may be up to 100 potential oncogenes in the human genome (about 0.2% of active genes). Although we undoubtedly still have a lot to learn from RNA tumour viruses and the biology of viral transformation, the emphasis has definitely shifted towards the analysis of proto-oncogenes in human tumour cells.

We still know very little about the expression of proto-oncogenes in a wide variety of normal cells and as yet no striking conclusions have been reached by looking at expression in tumour cells. The analysis of proto-oncogene organization in tumour cells has been much more rewarding. Genetic alterations at the *c-myc*, *c-abl* and *N-myc* loci are clearly important events in the development of Burkitt's lymphomas, chronic myeloid leukaemias and neuroblastomas. The transfection assay, however, has probably had the single greatest impact on the molecular biology of human cancer. The discovery that the three *ras* genes have undergone somatic mutation in many tumour cells has left little doubt that these genes are involved in a wide variety of human malignancies.

It should be clear from section V that we know very little about how oncogene proteins function in cells and the next five years will see a greater emphasis placed on this aspect of transformation. The oncogene products, and in particular the tyrosine kinases and the ras proteins, have stimulated research not only into the basis of transformation but also into the biology and physiology of the normal cell. The two areas of research have merged, results obtained in one being intimately linked to observations in the other. This might not be too surprising: cells divide in response to external stimuli which interact with membrane receptors. These receptors then generate intracellular second messengers which activate cytoplasmic and nuclear proteins, resulting in cell division and growth. A breakdown in control at any point in this cascade could be imagined to have lethal effects on a cell. Much more restricted perhaps might be alterations which are not lethal but which lead to an uncoupling of the controlling mechanisms at play. These represent the sites of action of possible proto-oncogenes in normal cells which by genetic alteration can give rise to oncogenes. An accumulation of several of these events, each affecting different aspects of the cell's physiology, could lead to the development of a cancer cell.

The goal of cancer research is to be able to diagnose and eventually to prevent or cure the disease. An understanding of which genes are important in which cancers might help in diagnosing premalignant stages of the disease. The mechanism by which proto-oncogenes are converted into oncogenes might help to point to ways in which at least some types of malignant disease can be avoided but the cure of cancer will first require a much deeper appreciation of the biochemical properties of oncogene proteins and how they interact with the cell.

VII   Acknowledgements

I should like to thank Robin Brown, Chris Marshall, Ian McKay and Hugh Paterson for help during the preparation of this manuscript. I should also like to thank the Institute of Cancer Research, the Cancer Research Campaign and the Medical Research Council for supporting the work on oncogenes being carried out in my laboratory.

VIII   References

Abrams, H. D., Rohrschneider, L. R. and Eisenman, R. N. (1982). *Cell* **29**, 427–439.
Anzano, M. A., Roberts, A. B., Smith, J. M., Sporn, M. B. and De Larco, J. E. (1983). *Proc. natn. Acad. Sci. USA* **80,** 6264–6268.
Armelin, H. A., Armelin, M. C. S., Kelly, K., Stewart, T., Leder, P., Cochran, B. H. and Stiles, C. D. (1984). *Nature, Lond.* **310,** 655–660.
Balmain, A., Ramsden, M., Bowden, G. T. and Smith, J. (1984). *Nature, Lond.* **307,** 658–660.
Baltimore, D. (1970). *Nature, Lond.* **226,** 1209–1211.
Barnekow, A. and Bauer, H. (1984). *Biochim. Biophys. Acta* **782,** 94–102.

Barnes, D. and Sato, G. (1980). *Anal. Biochem.* **102**, 255–270.

Barrett, T. B., Gajdnsek, C. M., Schwartz, S. M., McDougall, J. K. and Benditt, E. P. (1984). *Proc. natn. Acad. Sci. USA* **81**, 6772–6774.

Battey, J., Moulding, C., Taub, R., Murphy, W., Stewart, T., Potter, H., Lenoir, G. and Leder, P. (1983). *Cell* **34**, 779–787.

Bechade, C., Calothy, G., Pessac, B., Martin, P., Coll, J., Denhez, F., Saule, S., Ghysdael, J. and Stehelin, D. (1985). *Nature, Lond.* **316**, 559–562.

Benedict, W. F., Murphree, A. L., Banerjee, A., Spina, C. A. and Sparkes, M. C. (1983). *Science, N.Y.* **219**, 973–975.

Bernstein, A., MacCormick, R. and Martin, G. S. (1976). *Virology* **70**, 206–209.

Bishop, J. M. (1983). *Ann. Rev. Biochem.* **52**, 301–354.

Blair, D. G., Oskarsson, M., Wood, T. G., McClements, W. L., Fishchinger, P. J. and Vande Woude, G. F. (1981). *Science, N.Y.* **212**, 941–943.

Blenis, J., Spivack, J. G. and Erikson, R. L. (1984). *Proc. natn. Acad. Sci. USA* **81**, 6408–6412.

Bos, J. L., Verlaan-de Vries, M., Jansen, A. M., Veeneman, G. M., Van Boom, J. H. and van der Eb, A. J. (1984). *Nucl. Acids Res.* **12**, 9155–9163.

Bos, J. L., Toksoz, D., Marshall, C. J., Vries, M. V., Veeneman, G. H., van der Eb, A. J., van Boom, J. H., Janssen, J. W. G. and Steenvoorden, A. C. M. (1985). *Nature, Lond.* **315**, 726–730.

Brugge, J.S., Cotton, P. S., Queral, A. E., Barrett, J. N., Nonner, D. and Keane, R. W. (1985). *Nature, Lond.* **316**, 554–557.

Cairns, J. (1978). *"Cancer, Science and Society"*. W. H. Freeman, San Francisco, CA.

Campisi, J., Gray, H. E., Pardee, A. B., Dean, M. and Sonenshein, G. E. (1984). *Cell* **36**, 241–247.

Chang, E. H., Furth, M. E., Scolnick, E. M. and Lowy, D. R. (1982a). *Nature, Lond.* **297**, 479–483.

Chang, E. H., Gonda, M. A., Ellis, R. W., Scolnick, E. M. and Lowy, D. R. (1982b). *Proc. natn. Acad. Sci. USA* **79**, 4848–4852.

Clarke, M. F., Westin, E., Schmidt, D., Josephs, S. F., Ratner, L., Wong-Staal, F., Gallo, R. C. and Reitz, M. S. (1984). *Nature, Lond.* **308**, 464–467.

Coffin, J. M. and Billeter, M. A. (1976). *J. molec. Biol.* **100**, 293–318.

Cohen, S., Carpenter, G. and King, L. E. (1980). *J. biol. Chem.* **255**, 4834–4842.

Coll, J., Righi, M., De Taisne, C., Dissons, C., Gegonne, A. and Stehelin, D. (1983). *EMBO J.* **2**, 2189–2194.

Collett, M. S. and Erikson, R. L. (1978). *Proc. natn. Acad. Sci. USA* **75**, 2021–2024.

Collett, M. S., Erikson, E., Purchio, A. F., Brugge, J. S. and Erikson, R. L. (1979). *Proc. natn. Acad. Sci. USA* **76**, 3159–3163.

Collins, S. and Groudine, M. (1982). *Nature, Lond.* **298**, 679–681.

Collins, T., Ginsburg, D., Boss, J. M., Orkin, S. H. and Pober, J. S. (1985). *Nature, Lond.* **316**, 748–750.

Cooper, G. M., Okenquist, S. and Silverman, L. (1980). *Nature, Lond.* **284**, 418–421.

Cooper, J. A., Bowen-Pope, D. F., Raines, E., Ross, R. and Hunter, T. (1982). *Cell* **31**, 263–273.

Cooper, J. A., Reiss, N. A., Schwartz, R. J. and Hunter, T. (1983). *Nature, Lond.* **302**, 218–223.

Cooper, C. S., Park, M., Blair, D. G., Tainsky, M. A., Huebner, K., Croce, C. M. and Vande Woude, G. F. (1984). *Nature, Lond.* **311**, 29–33.

Cooper, J. A., Sefton, B. M. and Hunter, T. (1984). *Molec. cell. Biol.* **4**, 30–37.

Courtneidge, S. A., Levinson, A. D. and Bishop, J. M. (1980). *Proc. natn. Acad. Sci. USA* **77**, 3783–3787.

Cowell, J. K. (1982). *Ann. Rev. Genet.* **16**, 21–59.

Croce, C. M., Isobe, M., Palumbo, A., Puck, J., Ming, J., Tweardy, D., Erikson, J., Davis, M. and Rovera, G. (1985) *Science, N.Y.* **227**, 1044–1047.

Curran, T., Peters, G., Van Beveren, C., Teich, N. M. and Verma, I. M. (1982). *J. Virol.* **44**, 674–682.

Debuire, B., Henry, C., Benaissa, M., Biserte, G., Claverie, J. M., Saule, S., Martin, P. and Stehelin, D. (1984). *Science, N.Y.* **224**, 1456–1459.

Decker, S. (1981). *Proc. natn. Acad. Sci. USA* **78**, 4112–4115.

Defeo-Jones, D., Scolnick, E. M., Koller, R. and Dhar, R. (1983). *Nature, Lond.* **306**, 707–709.

De Larco, J. E. and Todaro, G. J. (1978). *Proc. natn. Acad. Sci. USA* **75**, 4001–4005.

Derynck, R., Roberts, A. B., Winkler, M. E., Chen, E. Y. and Goeddel, D. V. (1984). *Cell* **38**, 287–297.

Diamond, A., Cooper, G. M., Ritz, J. and Lane, M. A. (1983). *Nature, Lond.* **305**, 112–116.

Downward, J., Yarden, Y., Mayes, E., Scrace, G., Totty, N., Stockwell, P., Ullrich, A., Schlessinger, J. and Waterfield, M. D. (1984). *Nature, Lond.* **307**, 521–527.

Ek, B., Westermark, B., Wasteson, A. and Heldin, C. H. (1982). *Nature, Lond.* **295**, 419–420.

Ellis, R. W., DeFeo, D., Shih, T. Y., Gonda, M. A., Young, H. A., Tsuchida, N., Lowy, D. R. and Scolnick, E. M. (1981). *Nature, Lond.* **292**, 506–511.

Emanuel, B. S., Selden, J. R., Chaganti, R. S. K., Jhanwar, S., Nowell, P. C. and Croce, C. M. (1984). *Proc. natn. Acad. Sci. USA* **81**, 2444–2446.

Eva, A., Robbins, K. C., Andersen, P. R., Srinivasan, A., Tronick, S., Reddy, E. P., Ellmore, N. W., Galen, A. T., Lautenberger, J. A., Papas, T. S., Westin, E. H., Wong-Staal, F., Gallo, R. C. and Aaronson, S. A. (1982). *Nature, Lond.* **295**, 116–119.

Flockhart, D. A. and Corbin, J. D. (1982). *CRC Crit. Rev. Biochem.* **12**, 133.

Folkman, J. and Moscona, A. (1978). *Nature, Lond.* **273**, 345–349.

Frykberg, L., Palmieri, S., Beug, H., Graf, T., Hayman, M. J. and Vennstrom, B. (1983). *Cell* **32**, 227–238.

Fung, Y. K. T., Lewis, W. G., Crittenden, L. B. and Kung, H. J. (1983). *Cell* **33**, 357–368.

Gale, R. P. and Canaani, E. (1984). *Proc. natn. Acad. Sci. USA* **81**, 5648–5652.

Gambke, C., Signer, E. and Moroni, C. (1984). *Nature, Lond.* **307**, 476–478.

Gazit, A., Igarashi, H., Chiu, I. M., Srinivasan, A., Yaniv, A., Tronick, S. R., Robbins, K. C. and Aaronson, S. A. (1984). *Cell* **39**, 89–97.

Gilman, A. G. (1984). *Cell* **36**, 577–579.

Goubin, G., Goldman, D. S., Luce, J., Neiman, P. E. and Cooper, G. M. (1983). *Nature, Lond.* **302**, 114–118.

Graf, T. and Beug, H. (1978). *Biochim. Biophys. Acta* **516**, 269–299.

Graham, F. L. and van der Eb, A. J. (1973). *Virology* **52**, 456–461.

Graziani, Y., Maller, J. L., Sugimoto, Y. and Erikson, R. L. (1984). *In "Cancer Cells"* (G. F. Vande Woude, A. J. Levine, W. C. Topp and J. D. Watson, eds), Vol. 2, pp.27–35. Cold Spring Harbor Laboratory, Cold Spring Harbor, NY.

Guerrero, I., Calzada, P., Mayer, A. and Pellicer, A. (1984). *Proc. natn. Acad. Sci. USA* **81,** 202–205.

Hall, A. (1984). Oxford Surveys on Eukaryotic Genes **1,** 111–144.

Hall, A., Marshall, C. J., Spurr, N. and Weiss, R. A. (1983). *Nature, Lond.* **303,** 396–400.

Hampe, A., Laprevotte, I., Galibert, F., Fedele, L. A. and Sherr, C. J. (1982). *Cell* **30,** 775–785.

Hampe, A., Gobet, M., Sherr, C. J. and Galibert, F. (1984). *Proc. natn. Acad. Sci. USA* **81,** 85–89.

Hanafusa, T., Wang, L. H., Anderson, S. M., Karess, R. E., Hayward, W. S. and Hanafusa, H. (1980). *Proc. natn. Acad. Sci. USA* **77,** 3009–3013.

Hanafusa, H., Iba, H., Takeya, T. and Cross, F. R. (1984). *In* "Cancer Cells" (G. F. Vande Woude, A. J. Levine, W. C. Topp and J. D. Watson, eds), Vol. 2, pp. 1–7. Cold Spring Harbor Laboratory, Cold Spring Harbor, NY.

Harvey, J. J. (1964). *Nature, Lond.* **204,** 1104–1105.

Hayday, A. C., Gillies, S. D., Saito, H., Wood, C., Wiman, K., Hayward, W. S. and Tonegawa, S. (1984). *Nature, Lond.* **307,** 334–340.

Hayflick, L. and Moorhead, P. (1961). *Exp. cell. Res.* **25,** 585–621.

Hayward, W. S., Neel, B. G. and Astrin, S. M. (1981). *Nature, Lond.* **290,** 475–480.

Heisterkamp, N., Stephenson, J. R., Groffen, J., Hansen, P. F., de Klein, A., Bartram, C. R. and Grosveld, G. (1983). *Nature, Lond.* **306,** 239–242.

Heisterkamp, N., Stam, K., Groffen, J., Klein, A. and Grosveld, G. (1985). *Nature, Lond.* **315,** 758–761.

Hihara, M., Yamamoto, H., Shimohira, H., Arai, K. and Shimizu, T. (1983). *J. natn. Cancer Inst.* **70,** 891–895.

Houweling, A., van den Elsen, P. J. and van der Eb, A. J. (1980). *Virology* **105,** 537–550.

Hsu, Y. M., Barry, J. M. and Wang, J. L. (1984). *Proc. natn. Acad. Sci. USA* **81,** 2107–2111.

Hunter, T. (1984). *Sci. Am.* **251(2),** 60–69.

Hunter, T. and Sefton, B. (1980). *Proc. natn. Acad. Sci. USA* **77,** 1311–1315.

Hurley, J. B., Simon, M. I., Teplow, D. B., Robishaw, J. D. and Gilman, A. G. (1984). *Science* **226,** 860–862.

Hynes, R. (1982). *Cell* **28,** 437–438.

Jakobovits, E. B., Majors, J. E. and Varmus, H. E. (1984). *Cell* **38,** 757–765.

Jansen, H. W., Lurz, R., Bister, K., Bonner, T. I., Mark, G. E. and Rapp, U. R. (1984). *Nature, Lond.* **307,** 281–284.

Johnsson, A., Heldin, C.- H., Wasteson, A., Westermark, B., Deuel, T., Huang, J. S., Seeburg, P. H., Gray, A., Ullrich, A., Scrace, G., Stroobant, P. and Waterfield, M. D. (1984). *EMBO J.* **3,** 921–928.

Jonak, G. J. and Knight, E. (1984). *Proc. natn. Acad. Sci. USA* **81,** 1747–1750.

Kahn, P., Adkins, B., Beug, H. and Graf, T. (1984). *Proc. natn. Acad. Sci. USA* **81,** 7122–7126.

Kan, N. C., Flordellis, C. S., Mark, G. E., Duesberg, P. H. and Papas, T. S. (1984). *Proc. natn. Acad. Sci. USA* **81,** 3000–3004.

Kasuga, M., Yamaguchi, Y. F., Blithe, D. L. and Kahn, C. R. (1983). *Proc. natn. Acad. Sci. USA* **80,** 2137–2141.

Kataoka, T., Powers, S., McGill, C., Fasano, O., Strathern, J., Broach, J. and Wigler, M. (1984). *Cell* **37,** 437–445.

Kelly, K., Cochran, B. H., Stiles, C. D. and Leder, P. (1983). *Cell* **35,** 603–610.

Kennedy, N., Knediltscheck, G., Groner, B., Hynes, N. E., Herrlich, P., Michalides, R. and Van Ooyen, A. (1982). *Nature, Lond.* **295**, 622–624.

Kirsten, W. H. and Mayer, L. A. (1967). *J. natn. Cancer Inst.* **39**, 311–335.

Kitamura, N., Kitamura, A., Toyoshima, K., Hirayama, Y. and Yoshida, M. (1982). *Nature, Lond.* **297**, 205–208.

Klein, G. (1983). *Cell* **32**, 311–315.

Krontiris, T. G and Cooper, G. M. (1981). *Proc. natn. Acad. Sci. USA* **78**, 1181–1184.

Lai, M. M. C., Hu, S. S. F. and Vogt, P. K. (1979). *Virology* **97**, 366.

Land, H., Parada, L. F. and Weinberg, R. A. (1983). *Nature, Lond.* **304**, 596–602.

Lane, M. A., Sainten, A. and Cooper, G. M. (1982). *Cell* **28**, 873–880.

Lane, M. A., Sainten, A., Doherty, K. M. and Cooper, G. M. (1984). *Proc. natn. Acad. Sci. USA* **81**, 2227–2231.

Leder, P., Battey, J., Lenoir, G., Moulding, C., Murphy, W., Potter, H., Stewart, T. and Taub, R. (1983). *Science, N.Y.* **222**, 765–771.

Lee, F., Mulligan, R., Berg, P. and Ringold, G. M. (1981). *Nature, Lond.* **294**, 228–232.

Leprince, D., Gegonne, A., Coll, J., de Taisne, C., Schneeberger, A., Lagron, C. and Stehelin, D. (1983). *Nature, Lond.* **306**, 395–397.

Libermann, T. A., Nusbaum, H. R., Razan, N., Kris, R., Lax, I., Soreq, H., Whittle, N., Waterfield, M. D., Uhlrich, A. and Schlessinger, J. (1985). *Nature, Lond.* **313**, 144–146.

Linial, M. (1982). *Virology* **119**, 382–391.

Macara, I. G., Marinetti, G. V. and Balduzzi, P. C. (1984). *Proc. natn. Acad. Sci. USA* **81**, 2728–2732.

Macpherson, I. and Montagnier, L. (1964). *Virology* **23**, 291–294.

Manaker, R. A. and Groupe, V. (1956). *Virology* **2**, 838–840.

Marshall, C. J. and Rigby, P. W. J. (1984). *Cancer Surv.* **3**, 183–214.

Martin, G. S. (1970). *Nature, Lond.* **227**, 1021–1023.

McGrath, J. P., Capon, D. J., Goeddel, D. V. and Levinson, A. D. (1984). *Nature, Lond.* **310**, 644–649.

Moelling, K., Heimann, B., Beimling, P., Rapp, U. R. and Sander, T. (1984). *Nature, Lond.* **312**, 558–561.

Moscovici, C. and Zanetti, M. (1970). *Virology* **42**, 61–67.

Mulcahy, L. S., Smith, M. R. and Stacey, D. W. (1985). *Nature, Lond.* **313**, 241–243.

Muller, R., Bravo, R., Burckhardt, J. and Curran, T. (1984). *Nature, Lond.* **312**, 716–720.

Murray, M. J., Cunningham, J. M., Parada, L. F., Dautry, F., Lebowitz, P. and Weinberg, R. A. (1983). *Cell* **33**, 749–757.

Neharro, G., Robbins, K. C. and Reddy, E. P. (1984). *Science, N.Y.* **223**, 63–66.

Neckaneyer, W. S. and Wang, L. H. (1984). *J. Virol.* **50**, 914–921.

Newbold, R. F., Overell, R. W. and Connell, J. R. (1982). *Nature, Lond.* **299**, 633–635.

Nishikura, K., ar-Rushdi, A., Erikson, J., Watt, R., Govera, G. and Croce, C. M. (1983). *Proc. natn. Acad. Sci. USA* **80**, 4822–4826.

Nishizuka, Y. (1984a). *Nature, Lond.* **308**, 693–697.

Nishizuka, Y. (1984b). *Science, N.Y.* **225**, 1365–1369.

Nusse, R. and Varmus, H. E. (1982). *Cell* **31**, 99–109.

Palmieri, S., Kahn, P. and Graf, T. (1983). *EMBO J.* **2**, 2385–2389.

Parada, L. F., Land, H., Weinberg, R. A., Wolf, D. and Rotter, V. (1984). *Nature, Lond.* **312**, 649–651.

Parker, R. C., Varmus, H. E. and Bishop, J. M. (1984). *Cell* **37**, 131–139.

Payne, G. S., Bishop, J. M. and Varmus, H. E. (1982). *Nature, Lond.* **295**, 209–214.

Peters, G., Brookes, S., Smith, R. and Dickson, C. (1983). *Cell* **33**, 369–377.

Poiesz, B. J., Ruscetti, F. W., Gazder, A. F., Bunn, P. A., Minna, J. D. and Gallo, R. C. (1980). *Proc. natn. Acad. Sci. USA* **77**, 7415–7419.

Ponten, J. (1970). *Int. J. Cancer* **6**, 323–332.

Pope, J. H., Scott, W. and Moss, D. J. (1973). *Nature New Biol.* **246**, 140–141.

Popovic, M., Sarin, P. S., Robert-Guroff, M., Kalyanoraman, V. S., Mann, D., Minowada, J. and Gallo, R. C. (1983). *Science, N.Y.* **219**, 856–859.

Powers, S., Kataoka, T., Fasano, O., Goldfarb, M., Strathern, J., Broach, J. and Wigler, M. (1984). *Cell* **36**, 607–612.

Rabbits, T. H., Hamlyn, P. H. and Baer, R. (1983). *Nature, Lond.* **306**, 760–765.

Radke, K., Beug, H., Kornfed, S. and Graf, T. (1982). *Cell* **31**, 643–653.

Ramsey, G., Graf, T. and Hayman, M. J. (1980). *Nature, Lond.* **288**, 170–172.

Rapp, U. R., Goldsborough, M. D., Mark, G. E., Bonner, T. I., Groffen, J., Reynolds, F. H. and Stephenson, J. R. (1983). *Proc. natn. Acad. Sci. USA* **80**, 4218–4222.

Rassoulzadegan, M., Cowie, A., Carr, A., Glaichenhaus, N., Kamen, R. and Cuzin, F. (1982). *Nature, Lond.* **300**, 713–718.

Rassoulzadegan, M., Naghashfar, Z., Cowie, A., Carr. A., Grisoni, M., Kamen, R. and Cuzin, F. (1983). *Proc. natn. Acad. Sci. USA* **80**, 4354–4358.

Ratner, L., Josephs, S. F., Jarrett, R., Reitz, M. S. and Wong-Staal, F. (1985). *Nucl. Acids. Res.* **13**, 5007–5018.

Reddy, E. P., Reynolds, R. K., Santos, E. and Barbacid, M. (1982). *Nature, Lond.* **300**, 149–152.

Reddy, E. P., Smith, M. J. and Srinivasan, A. (1983). *Proc. natn. Acad. Sci. USA* **80**, 3623–3627.

Rodriguez-Pena, A. and Rozengurt, E. (1985). *EMBO J.* **4**, 71–76.

Rous, P. (1911). *J. exp. Med.* **13**, 397–411.

Rowley, J. D. (1980). *Ann. Rev. Genet.* **14**, 17–39.

Royer-Pokora, B., Beug, H., Claviez, M., Winkhardt, H. J., Friis, R. R. and Graf, T. (1978). *Cell* **13**, 751–760.

Ruley, E. H. (1983). *Nature, Lond.* **304**, 602–606.

Rushdi, A., Nishikura, K., Erikson, J., Watt, R., Rovera, G. and Croce, C. M. (1983). *Science, N.Y.* **222**, 390–393.

Rushlow, K. E., Lautenberger, J. A., Papas, T. S., Baluda, M. A., Perbal, B., Chirikjian, J. G. and Reddy, E. P. (1982). *Science, N.Y.* **216**, 1421.

Saito, H., Hayday, A. C., Wiman, K., Hayward, W. S. and Tonegawa, S. (1983). *Proc. natn. Acad. Sci. USA* **80**, 7476–7480.

Santos, E., Tronick, S. R., Aaronson, S. A., Pulciani, S. and Barbacid, M. (1982). *Nature, Lond.* **298**, 343–347.

Schecter, A. L., Stern, F. F., Vaidyanathan, L., Decker, S. J., Drebin, J. A., Greene, M. I. and Weinberg, R. A. (1984). *Nature, Lond.* **312**, 513–516.

Schimke, R. T. (ed.) (1982). "Gene Amplification". Cold Spring Harbor Laboratory, Cold Spring Harbor, N.Y.

Schimke, R. T., Kaufman, R. J., Alt, F. W. and Kellems, R. F. (1978). *Science, N.Y.* **202**, 1051–1055.

Schwab, M., Alitalo, K., Klempnauer, K. H., Varmus, H. E., Bishop, J. M.,

Gilbert, F., Brodeur, G., Goldstein, M. and Trent, J. (1983). *Nature, Lond.* **305**, 245–247.

Schwab, M., Alitalo, K., Varmus, H. E. and Bishop, J. M. (1984). In "Cancer Cells" (G. F. Vande Woude, A. J. Levine, W. C. Topp and J. D. Watson, eds), Vol. 2, pp.215–220. Cold Spring Harbor Laboratory, Cold Spring Harbor, N.Y.

Schwartz, D. E., Tizard, R. and Gilbert, W. (1983). *Cell* **32**, 853–869.

Seiki, M., Hattori, S., Hirayama, Y. and Yoshida, M. (1983). *Proc. natn. Acad. Sci. USA* **80**, 3618–3622.

Seiki, M., Eddy, R., Shows, T. B. and Yoshida, M. (1984). *Nature, Lond.* **309**, 640–642.

Shalloway, D., Coussens, P. M. and Yaciuk, P. (1984). *Proc. natn. Acad. Sci. USA* **81**, 7071–7075.

Sherr, C. J., Rettenmier, C. W., Sacca, R., Roussel, M. F., Look, A. T. and Stanley, E. R. (1985). *Cell* **41**, 665–676.

Shih, C. and Weinberg, R. A. (1982). *Cell* **29**, 161–169.

Shih, C., Shilo, B. Z., Goldfarb, M. P., Dannenberg, A. and Weinberg, R. A. (1979). *Proc. natn. Acad. Sci. USA* **76**, 5714–5718.

Shih, C., Padhy, L. C., Murray, M. and Weinberg, R. A. (1981). *Nature, Lond.* **290**, 261–264.

Shih, C. K., Linial, M., Goodenow, M. M. and Hayward, W. S. (1984). *Proc. natn. Acad. Sci. USA* **81**, 4697–4701.

Shilo, B. Z. and Weinberg, R. A. (1981). *Proc. natn. Acad. Sci. USA* **78**, 6789–6792.

Shimizu, K., Goldfarb, M., Perucho, M. and Wigler, M. (1983). *Proc. natn. Acad. Sci. USA* **80**, 383–387.

Slamon, D. J., de Kernion, J. B., Verma, I. M. and Cline, M. J. (1984a). *Science, N.Y.* **224**, 256–262.

Slamon, D. J., Shimotohno, K., Cline, M., Golde, D. W. and Chen, I. S. Y. (1984b). *Science, N.Y.* **226**, 61–64.

Sodroski, J. G., Rosen, C. A. and Haseltine, W. A. (1984). *Science, N.Y.* **225**, 381–385.

Spandidos, D. and Wilkie, N. M. (1984). *Nature, Lond.* **310**, 469–475.

Spector, D. H., Varmus, H. E. and Bishop, J. M. (1978). *Proc. natn. Acad. Sci. USA* **75**, 4102–4106.

Sporn, M. B. and Todaro, G. J. (1980). *N. Engl. J. Med.* **303**, 878–880.

Stavnezer, E., Gerhard, D. S., Binari, R. C. and Balazs, T. (1981). *J. Virol.* **39**, 920–934.

Stehelin, D., Varmus, H. E., Bishop, J. M. and Vogt, P. K. (1976). *Nature, Lond.* **260**, 170–173.

Stiles, C. D. (1983). *Cell* **33**, 653–655.

Sugimoto, Y., Whitman, M., Cantley, L. C. and Erikson, R. L. (1984). *Proc. natn. Acad. Sci. USA* **81**, 2117–2121.

Sushilkumar, G., Devare, E., Reddy, E. P., Law, J. D., Robbins, K. C. and Aaronson, S. A. (1983). *Proc. natn. Acad. Sci. USA* **80**, 731–735.

Tabin, C. J., Bradley, S. M., Bargmann, C. I., Weinberg, R. A., Papageorge, A. G., Scolnick, E. M., Dhar, R., Lowy, D. R. and Chang, E. H. (1982). Nature, Lond. **300**, 143–149.

Takeya, T. and Hanafusa, H. (1982). *J. Virol.* **44**, 12–18.

Temin, H. M. and Mizutani, S. (1970). *Nature, Lond.* **226**, 1211–1213.

Toda, T., Uno, I., Ishikawa, T., Powers, S., Kataoka, T., Brock, D., Cameron, S., Broach, J., Matsumoto, K. and Wigler, M. (1985). *Cell* **40**, 27–36.

Todaro, G. J., Fryling, C. and De Larco, J. E. (1980). *Proc. natn. Acad. Sci. USA* **77,** 5258–5262.

Tonegawa, S. (1983). *Nature, Lond.* **302,** 575–581.

Tooze, J. (1981). "Molecular Biology of Tumor Viruses. DNA Tumor Viruses" 2 edn. Cold Spring Harbor Laboratory, Cold Spring Harbor, N.Y.

Toyoshima, K. and Vogt, P. K. (1969). *Virology* **39,** 930–931.

Tsujimoto, Y., Jaffe, E., Cossman, J., Gorham, J., Nowell, P. C. and Croce, C. M. (1985). *Nature, Lond.* **315,** 340–343.

Twardzik, D. R., Ranchalis, J. E. and Todaro, G. J. (1982). *Cancer Res.* **42,** 590–593.

Van Beveren, C., Van Straaten, F., Galleshaw, J. A. and Verma, I. M. (1981). *Cell* **27,** 97–108.

Van Beveren, C., Van Straaten, F., Curran, T., Muller, R. and Verma, I. M. (1983). *Cell* **32,** 1241–1255.

Van Beveren, C., Gnami, S., Curran, T. and Verma, I. M. (1984). *Virology* **135,** 229–243.

Van Ooyen, A. and Nusse, R. (1984). *Cell* **39,** 233–240.

Vennstrom, B., Kahn, P., Adkins, B., Enrietto, P., Hayman, M. J., Graf, T. and Luciw, P. (1984). *EMBO J.* **3,** 3223–3229.

Vousden, K. H. and Marshall, C. J. (1984). *EMBO J.* **3,** 913–917.

Wang, L. H., Duesberg, P., Beeman, K. and Vogt, P. K. (1975). *J. Virol.* **161,** 1051–1070.

Waterfield, M. D., Scrace, G. T., Whittle, N., Stroobant, P., Johnson, A., Wasteson, A., Westermark, B., Heldin, C. H., Huang, J. S. and Deuel, T. F. (1983). *Nature, Lond.* **304,** 35–39.

Watson, D. K., Psallidopoulos, M. C., Samuel, K. P., Dalla-Favera, R. and Papas, T. S. (1983). *Proc. natn. Acad. Sci. USA* **80,** 3642–3645.

Watt, R., Stanton, L. W., Marcu, K. B., Gallo, R. C., Croce, C. M. and Rovera, G. (1983). *Nature, Lond.* **303,** 725–728.

Weiss, R. (1970). *Int. J. Cancer* **6,** 333–345.

Weiss, R. A. (1973). *In* "Possible Episomes in Eukaryotes" (L. G. Silvestri, ed.) pp. 130–141. North-Holland, Amsterdam.

Weiss, R. A., Teich, N., Varmus, H. and Coffin, J. (1985). *In* "Molecular Biology of Tumor Viruses. RNA Tumor Viruses" 2nd edn. Cold Spring Harbor Laboratory, Cold Spring Harbor, N.Y.

Westin, E. M., Gallo, R. C., Arya, S. K., Eva, A., Souza, L. M., Baluda, M. A., Aaronson, S. A. and Wong-Staal, F. (1982a). *Proc. natn. Acad. Sci. USA* **79,** 2194–2198.

Westin, E. H., Wong-Staal, F., Gelman, E. P., Dalla Favera, R., Papas, T. S., Lautenberger, J. A., Eva, A., Reddy, E. P., Tronick, S. R., Aaronson, S. A. and Gallo, R. C. (1982b). *Proc. natn. Acad. Sci. USA* **79,** 2490–2494.

Wood, T. G., McCready, M. L., Blair, D. G. and Vande Woude, G. F. (1983). *J. Virol.* **46,** 726–736.

Wyke, J. (1983). *Nature, Lond.* **304,** 491–492.

Zarbl, H., Sukumar, S., Arthur, A. V., Martin-Zanca, D. and Barbacid, M. (1985). *Nature, Lond.* **315,** 382–385.

Genes of the immune system

MICHAEL STEINMETZ

Basel Institute for Immunology, Grenzacherstrasse 487, CH-4005 Basel, Switzerland

I	Introduction	118
II	Major histocompatibility complex	119
	A Structure and function of major histocompatibility complex class I and class II molecules	119
	B Genetic map	121
III	Immunoglobulin genes	123
	A Structure of immunoglobulin molecules	124
	B The Dreyer–Bennett hypothesis	125
	C Organization and rearrangement of immunoglobulin genes. .	129
IV	T-cell receptor genes	130
	A The T-cell receptor complex	132
	B Organization and rearrangement of T-cell receptor α and β chain genes	133
	C The T-cell-receptor-related γ chain gene	133
V	The immunoglobulin superfamily	134
	A Sequence relationships	134
	B The immunoglobulin homology unit	136
	C Rearranging immunoglobulin and T-cell receptor gene segments	138
	D Chromosomal locations and translocations	138
VI	Generation of diversity	141
	A Major histocompatibility complex alleles	138
	B Immunoglobulin and T-cell receptor genes	141
VII	Species comparison	145
	A Major histocompatibility complex genes	145
	B Immunoglobulin genes	146
	C T-cell receptor genes	147
VIII	Conclusions	147
IX	Acknowledgements	148
X	References	148

Genetic Engineering Vol. 5
ISBN 0-12-270305-7

I Introduction

Living organisms have developed several systems which they use to communicate with the outside world. These systems are capable of receiving external signals, of transferring information to other molecules or cells inside the organism and of responding in specific ways. The vertebrate immune system has all these properties. It can detect foreign molecular entities, transfer information between different structural elements of the system and respond by producing cells and molecules which will finally lead to the removal of foreign matter. The function of the immune system is to defend the organism against foreign invaders. It fulfils its function by tolerating self-derminants (these are all the determinants which constitute the organism itself) and responding to foreign, or non-self-determinants (the outside world).

The immune system is composed of lymphatic organs, such as the thymus, spleen, bone marrow and lymph nodes, and lymphocytes, which travel as single cells throughout the body (except the brain) (Klein, 1982; Paul, 1984). Lymphocytes, which can be divided into B lymphocytes and T lymphocytes, are concerned with the distinction between self and non-self. To carry out their function B and T lymphocytes synthesize certain receptor molecules which are expressed on the cell surface and can bind to other molecules, collectively called antigens. In the case of B lymphocytes, the antigen receptor is the immunoglobulin molecule which can be synthesized in two different forms: either as a membrane-bound molecule or as a molecule secreted by the cell. The antigen receptor of T lymphocytes has been termed the T-cell receptor and is usually found as a membrane-bound molecule.

B and T lymphocytes recognize antigen in two different ways. Whereas immunoglobulin molecules, synthesized by B cells, can bind soluble antigen directly, the T-cell receptor cannot. T cells need the antigen presented to them on the surface of an accessory cell and recognize it in conjunction with certain membrane molecules. The molecules which are recognized together with foreign antigen are the class I and class II molecules of the major histocompatibility complex (MHC). The detection of foreign antigens in the context of MHC molecules has been termed MHC restriction, although the precise molecular interactions between the three components (T-cell receptor, MHC molecule and foreign antigen) are not understood. Two hypotheses have been put forward. One postulates that T cells have two receptors, one for foreign antigens and one for the MHC molecule (dual recognition). The second assumes that T cells use one receptor to detect a complex of MHC molecule and foreign antigen (Schwartz, 1985).

Isolation and characterization of the genes encoding immunoglobulin, MHC and T-cell receptor molecules have shown that they are all related and are probably derived from a common ancestral gene. It appears that this primordial gene encoded an immunoglobulin-like domain which was capable of interacting with itself and with other protein domains. Today the important feature of this domain is that it can accommodate, within its structure, a large variety of different amino acid sequences (Hood *et al.*, 1985). These variable positions in the immunoglobulin-like domain are called hypervariable regions. Since the hypervariable regions form part of the antigen-binding site, immunoglobulin-like domains can be generated which exhibit exquisite binding specificity.

In order to recognize a large number of foreign antigens the immune system has developed several mechanisms to generate, within the lifetime of an individual, a large repertoire of distinct T-cell receptor and antibody molecules. Extreme variability is also observed in MHC molecules, although not within the individual but between different members of a species.

In this chapter I will discuss the structure, function and organization of the MHC, immunoglobulin and T-cell receptor genes of the mouse; I will look at the different mechanisms which create diversity in the three gene families and briefly compare the murine immune system genes with those of other species.

II   Major histocompatibility complex

A   Structure and function of major histocompatibility complex class I and class II molecules

The MHC in the mouse is historically defined as a large gene cluster on chromosome 17 coding for molecules which induce rapid tissue graft rejections between genetically distinct animals (Gorer, 1936; Snell, 1981). According to current thinking it would be better to define the MHC as a gene cluster coding for T-cell guidance molecules which "restrict" the specificity of T cells (Klein *et al.*, 1983b; Schwartz, 1984b). In the BALB/c mouse strain five different cell surface glycoproteins have been identified which possess this function. These are the class I molecules K, D and L, and the class II molecules I-A and I-E. The three class I molecules restrict the specificity of cytotoxic T cells and the two class II molecules are recognized by helper T cells together with foreign antigen. The class I and class II molecules acting as T-cell restriction elements differ extensively between genetically distinct mouse strains; in other words, they exhibit a great deal of allelic variability (see below). For that reason, T cells from one individual, recognizing a given antigen together with a certain MHC molecule, will in general not recognize the same antigen presented in the context of an MHC molecule encoded by the allelle of a different individual. Instead, T cells show a high degree of alloreactivity towards foreign MHC molecules which they presumably mistake for an altered self MHC molecule. This explains why T cells reject tissue grafts displaying foreign MHC molecules.

Class I and class II molecules are each composed of two polypeptide chains (Fig. 1). Whereas both the α and the β chains of class II molecules are encoded in the MHC on chromosome 17, only the α chain of class I molecules is. The gene coding for β_2-microglobulin, the light chain of class I molecules, is located on chromosome 2 in the mouse. Both class I

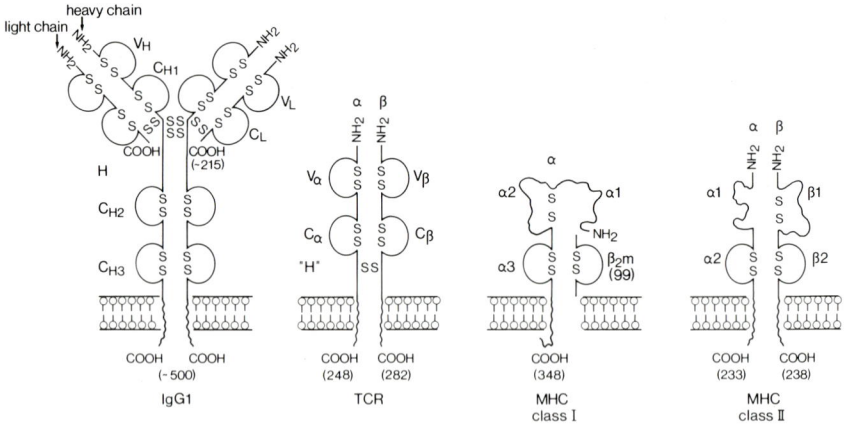

Figure 1   Domain organization of immunoglobulin, T-cell receptor and MHC class I and class II molecules. Immunoglobulin-like domains, related to one another because of significant amino acid homology, are indicated by disulphide-bonded protein loops. Domains of class I and class II molecules which exhibit pronounced allelic variability are drawn as irregular shaped loops. Domains are indicated by V_H, C_H, V_α, V_β, H, α1, β1 etc. (see text). The lengths of polypeptide chains are given in parentheses and were taken from the following sources: immunoglobulin G1 (IgG1), Kabat *et al.* (1983); T-cell receptor (TCR) Saito *et al.* (1984a, 1984b); MHC class I and class II (K^d and I-A^d respectively), Steinmetz and Hood (1983); β_2-microglobulin (β_2m), Parnes and Seidman (1982). The size difference observed between α and β chains of the T-cell receptor is mainly due to a larger C_β domain and larger hinge-like region (H) (Gascoigne *et al.*, 1984) of the β chain (Chien *et al.*, 1984b; Saito *et al.*, 1984a, 1984b).

and class II molecules consist of four extracellular protein domains, a transmembrane region and small cytoplasmic portion. The variability between different alleles is mainly confined to the two amino-terminal domains (α1 and α2 for class I, α1 and β1 for class II) whereas the membrane-proximal domains are conserved and show significant homology to immunoglobulin domains (see below) (Hood *et al.*, 1983; Steinmetz and Hood, 1983; Mengle-Gaw and McDevitt, 1985; Steinmetz, 1986).

B   Genetic map

As is evident from Fig. 2, the MHC is an enormous genetic region containing at least 46 genes in the BALB/c mouse, namely 33 class I genes, seven class II genes and six genes unrelated to class I or class II. On the basis of recombination frequencies the MHC is known to cover about 2 centimorgans or about 1/30 of chromosome 17 (Klein *et al.*, 1982). This would correspond to about 4000 kb if there is a linear relationship between genetic length, measured in centimorgans, and physical length, measured in base pairs, which, however, is not the case as will

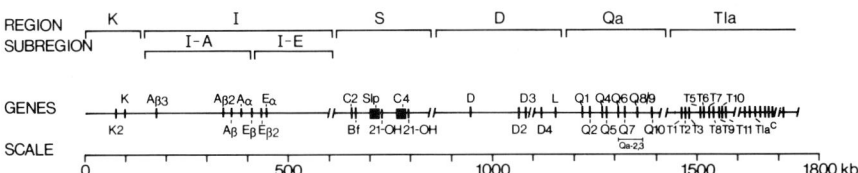

Figure 2   The MHC of the BALB/c mouse. Genes are indicated by boxes. Class I genes in the Qa and Tla regions are labelled using the nomenclature established by Weiss *et al.* (1984) for C57BL/10 class I genes. Unlabelled BALB/c class I genes in the Tla region do not correspond to class I genes cloned from C57BL/10 DNA. Q7 is identical with the previously sequenced class I gene 27.1 (Steinmetz *et al.*, 1981). The Qa-2, 3 antigen is encoded by the Q6, Q7 or Q8/9 gene, most probably the Q6 gene (Flaherty *et al.*, 1985; Mellor *et al.*, 1985). The TL antigen is encoded by the Tla gene (Goodenow *et al.*, 1982). The seven cloned portions encompassing part of the K and the I region, the D and Qa region and parts of the S, D, Qa and Tla regions are drawn to scale. The amount of DNA located between the cloned segments and the orientation of DNA segments which are located entirely within one region are not known. Also unknown is the order of the multiple segments in the Tla region. The maps of the seven cloned DNA segments are based on published and unpublished data from Steinmetz *et al.* (1982a, 1982b, 1984, 1986), Winoto *et al.* (1983), Chaplin *et al.* (1983), Amor *et al.* (1985), Fisher *et al.* (1985), Hämmerling *et al.* (1985) and Rogers (1985).

become evident later. For these reasons we have currently no good estimate for the true molecular size of the MHC. However, on the basis of the amount of DNA that has been cloned so far, we know that the MHC must be larger than 1700 kb.

Using a battery of specific antisera and monoclonal antibodies as a tool, the genes encoding different MHC molecules have been placed into six regions, called K, I, S, D, Qa and Tla. The boundaries of these regions are defined by crossing-over events which have occurred in certain recombinant mouse strains (Klein *et al.*, 1983a).

The class I and class II genes coding for T-cell restriction elements are located in the K, I and D regions. In the BALB/c mouse, this group of genes comprises at least three class I genes K, D and L, coding for the α chain of the K, D and L molecules, and four class II genes A_α, A_β, E_α and E_β, coding for the α and β polypeptides of the I-A and I-E molecules (Hood *et al.*, 1983; Steinmetz and Hood, 1983).

In addition to the genes coding for T-cell restriction elements, the MHC contains a number of other related and unrelated genes (Fig. 2). First, closely linked to the class I and class II genes encoding T-cell restriction elements, homologous genes have been identified, some of which appear to be at least transcriptionally active. In addition to the K, D and L genes there are four more class I genes located in the K and D regions of the BALB/c mouse (Steinmetz and Hood, 1983; Steinmetz *et al.*, unpublished results). They have not been characterized further. Also, three more class II β genes have been found in the I region (Steinmetz *et al.*, 1982a, 1984, 1986; Larhammar *et al.*, 1983; Widera and Flavell, 1985). The $A_{\beta3}$ gene is a pseudogene in at least two different mouse strains whereas the $A_{\beta2}$ and $E_{\beta2}$ genes, known to be transcribed at a low level in spleen cells, have no obvious mutation which would render them non-functional (Braunstein and Germain, 1985; Wake *et al.* 1985; Widera and Flavell, 1985).

Second, molecular cloning studies have revealed an unexpectedly large number of class I genes in the Qa and Tla regions (Steinmetz *et al.*, 1982b; Winoto *et al.*, 1983; Weiss *et al.*, 1984). In contrast with the K, D and L genes, which are expressed in virtually all somatic cells, at least some of the genes located in the Qa and Tla regions show tissue-specific expression (Flaherty, 1981; Michaelson *et al.*, 1983). Virtually nothing is known about the function of these genes. They do not encode molecules used by T cells as restriction elements and their characterization is currently under way in many laboratories. Genes have been identified which encode a liver-specific secreted class I molecule (Q10) (Mellor *et al.*, 1983; Maloy *et al.*, 1984; Lalanne *et al.*, 1985), a cell surface molecule found on haematopoietic cells (Qa-2) (Flaherty *et al.*, 1985; Mellor *et al.*,

1985) and a cell surface class I molecule expressed on thymocytes and certain T-cell leukaemias (Goodenow *et al.*, 1982; Boyse, 1984; Fisher *et al.*, 1985). At least two more class I molecules (Qa-1 and Mta) are encoded in the distal portion of the MHC but the corresponding genes have not yet been found (Flaherty, 1981; Fischer Lindahl *et al.*, 1983).

Third, six genes have been identified in the S region, three of which code for molecules (Bf, C2, C4) which are components of the classical and the alternative pathways of complement (Chaplin *et al.*, 1983; Steinmetz *et al.*, unpublished results). A fourth gene codes for a molecule called Slp which is closely related to C4 but has no haemolytic activity (Ogata and Sepich, 1984; Lévi-Strauss *et al.*, 1985). The remaining two genes, closely linked to Slp and C4, code for 21-hydroxylase, an enzyme involved in steroid biosynthesis (White *et al.*, 1984; Amor *et al.*, 1985). It is unclear why the genes in the S region are associated with class I and class II genes in essentially all the species that have been studied so far. Presumably the genes of the MHC represent a conserved linkage group which includes genes that have no functional relationship.

III   Immunoglobulin genes

A   Structure of immunoglobulin molecules

Three immunoglobulin gene families in the mouse code for three distinct families of polypeptide chains: two different families of light chains (κ and λ) and one family of heavy chains (Leder, 1982; Honjo, 1983; Max, 1984). A typical immunoglobulin molecule is composed of four polypeptide chains, two identical light chains and two identical heavy chains bound to each other by disulphide bridges (Fig. 1). A given antibody molecule has either two identical κ or two identical λ light chains. In the mouse, κ chains are found in 95% of the immunoglobulin molecules, whereas λ chains are found in only 5% of the molecules. λ chains are composed of three different types called $\lambda 1$, $\lambda 2$ and $\lambda 3$. The heavy chains are composed of eight different classes called μ, δ, $\gamma 1$, $\gamma 2a$, $\gamma 2b$, $\gamma 3$, ε and α. Again, a given antibody molecule will always have two chains of the same heavy chain class. For example, the IgG1 molecule consists of two $\gamma 1$ heavy chains and two light chains (either two λ or two κ light chains) (Fig. 1).

A number of structural and functional studies have indicated that antibody molecules are composed of protein domains with different recognition and effector functions. The IgG1 molecule shown in Fig. 1 consists of 14 domains, five domains for each heavy chain and two domains for each light chain. The heavy chain domains are denoted V_H, C_{H1}, H, C_{H2} and C_{H3}. V_H identifies a variable-region domain, C_{H1}, C_{H2} and C_{H3} identify three constant-region domains and H identifies the hinge region. The light chain is composed of a variable-region domain V_L and a constant-region domain C_L.

The distinction of variable and constant regions in light and heavy chains is based on the fact that different chains of the same light chain type or the same heavy chain class show amino acid sequence variation in their amino-terminal portions whereas the carboxy-terminal parts are constant. The structural division also reflects a functional division of the molecule: the variable regions of antibody molecules bind foreign antigens while the constant portions exert certain effector functions. The binding site for antigen is formed by both heavy and light chain variable regions and in particular by the so-called hypervariable regions, of which there are three in each variable region. These hypervariable regions are flanked by so-called framework regions which are more conserved. The framework residues are presumably important for the proper three-dimensional structure of the variable domain.

B   The Dreyer–Bennett hypothesis

Once it became known that antibodies specific for different antigens differed from each other in their variable regions but could share the same constant regions, the question arose as to how this information was encoded in the DNA. Two basic theories were debated. The first one postulated that the information to encode all the different antibodies is present in germline DNA. In contrast, the second one stated that only one gene for each class of antibody is present in germline DNA and that somatic mutation in antibody-producing cells would generate the sequence variation in the variable region.

There were problems with both of these models, however. In the case of the germline model, it seemed that too large a portion of the eukaryotic genome would be required to synthesize the millions of different antibody molecules. Furthermore, the absolute conservation of the constant region appeared difficult to explain since its multiple copies should undergo independent evolution and should show indications of genetic drift. Somatic theories for the generation of antibody diversity, however, made it difficult to explain the inheritance of certain genetic markers that had been discovered in the antigen-binding sites of the variable regions.

It was at this time of confusion that Dreyer and Bennett (1965) proposed their revolutionary hypothesis that variable and constant regions were encoded by separate genes in germline DNA. They assumed that for each heavy and light chain there was only a single constant-region gene whereas multiple genes encoding variable regions would be present in germline DNA. For the production of a complete heavy or light chain polypeptide, one of many variable-region genes would be fused to the single constant-region gene. This model, although revolutionary and therefore met with a lot of scepticism, turned out to be essentially correct. The fusion of the coding information for variable and constant regions occurs at the DNA level, and not at the RNA or protein level.

C Organization and rearrangement of immunoglobulin genes

1 λ *light chain genes*

The organization of the $\lambda 1$ light chain gene in germline DNA is compared with its organization in a B cell producing a $\lambda 1$ polypeptide in Fig. 3. In germline DNA the gene is split into four parts called $L_{\lambda 1}$, $V_{\lambda 1}$, $J_{\lambda 1}$ and $C_{\lambda 1}$ (Tonegawa *et al.*, 1977; Bernard *et al.*, 1978). The $L_{\lambda 1}$ exon encodes the leader or signal peptide which is required for proper intracellular transport of the $\lambda 1$ chain but is cleaved off from the mature molecule; the $V_{\lambda 1}$ gene segment encodes the first 98 amino acids of the variable region, the $J_{\lambda 1}$ gene segment encodes the remaining 12 amino acids of the variable region and the $C_{\lambda 1}$ exon the constant region of the $\lambda 1$ light chain. In the B cell a DNA rearrangement results in the fusion of the $V_{\lambda 1}$ and $J_{\lambda 1}$ (for joining) gene segments (Bernard *et al.*, 1978). This rearrangement allows the $\lambda 1$ light chain gene to be expressed. After transcription and splicing, the mRNA is translated into a $\lambda 1$ light chain polypeptide. Essentially similar DNA rearrangement and RNA splicing events are required for expression of κ light chain and heavy chain genes.

The mouse contains a total of four different C_λ exons which have been linked by molecular cloning into two gene clusters located on chromosome 16 (Blomberg *et al.*, 1981; Miller *et al.*, 1981) (Fig. 4). A single J gene segment is found a few thousand base pairs of DNA upstream of each C_λ exon. Only $C_{\lambda 3}$, $C_{\lambda 1}$ and $C_{\lambda 2}$, together with their corresponding J gene segments, are functional genes. $C_{\lambda 4}$ appears to be non-functional because of a mutated donor splice site in $J_{\lambda 4}$ and has never been found to be expressed. Two germline V_λ gene segments have been cloned and are

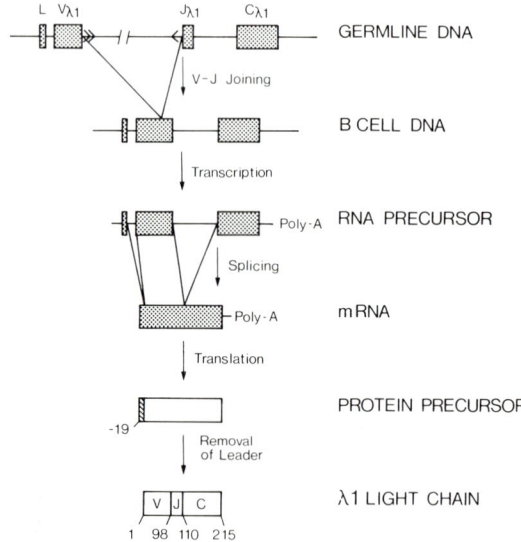

Figure 3 Rearrangement and expression of the λ1 light chain gene. Gene segments, RNA and protein molecules are not drawn to scale. The V gene segment contains an intron 93 bp long at codon position −4 of the leader peptide. The J gene segment is separated from the C region gene by an intron about 1.2 kb in size. Single or double arrow heads indicate one- or two-turn spacer signal sequences that are presumably important for the joining of gene segments. Numerals denote amino acid positions.

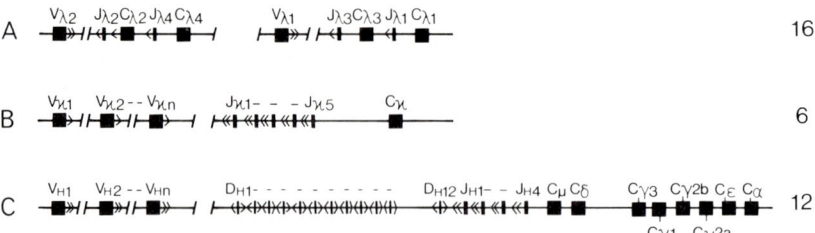

Figure 4 Organization of the three immunoglobulin gene families λ(A), κ(B) and heavy chain (C) genes, located on mouse chromosomes 16, 6 and 12 respectively. For explanation see the text and the legend to Fig. 3. The order and orientation of unlinked gene segments are not known. (Not drawn to scale.)

assumed to be linked to the two constant-region gene clusters in the way shown in Fig. 4, but a molecular linkage has not yet been established. The arrangement is based on the finding that, at the protein level, $V_{\lambda 2}$ is linked to $C_{\lambda 2}$, whereas $V_{\lambda 1}$ is linked to either $C_{\lambda 1}$ or $C_{\lambda 3}$. Furthermore, a hybridoma protein has been found with $V_{\lambda 2}$ linked to $C_{\lambda 3}$, suggesting that the $V_{\lambda 2}$–$C_{\lambda 2}$–$C_{\lambda 4}$ cluster lies upstream of $V_{\lambda 1}$–$C_{\lambda 3}$–$C_{\lambda 1}$ (Elliott *et al.*, 1982).

2   *κ light chain genes*

The κ chain gene family (Fig. 4) is located on chromosome 6 in the mouse. In contrast with the λ gene family it contains only a single C_κ exon but five J_κ gene segments, one of which ($J_{\kappa 3}$) appears to be non-functional because of an inappropriate donor splice site (Max *et al.*, 1979; Sakano *et al.*, 1979a). Furthermore, a large number of V_κ gene segments have been identified. Current estimates of the number of V_κ gene segments in the mouse range between 90 and 300 (Cory *et al.*, 1981). The distance as well as the orientation of the V_κ gene segments with respect to the C_κ exon are unknown. Joining of V_κ and J_κ gene segments is believed to be completely random, meaning that any V_κ gene segment could combine with any J_κ gene segment with equal probability.

3   *Heavy chain genes*

The organization of the immunoglobulin heavy chain genes, located on chromosome 12 in the mouse, is more complex than that of the light chain genes. Eight different constant-region genes, corresponding to the eight different classes of heavy chain, are located in a region of about 200 kb of DNA (Fig. 4) (Shimizu *et al.*, 1982). The order of the constant-region genes (in 5′-to-3′ orientation) is C_μ, C_δ, $C_{\gamma 3}$, $C_{\gamma 1}$, $C_{\gamma 2b}$, $C_{\gamma 2a}$, C_ε and C_α. Each constant-region gene is composed of a number of exons which correlate with distinct protein domains (Sakano *et al.*, 1979b). Upstream of the C_μ gene a cluster of four J_H gene segments has been identified, all of which are functional. In contrast with the light chain variable regions which are encoded by two gene segments, V and J, the variable regions of heavy chains are encoded by three gene segments, V_H, D_H and J_H (where D_H is the diversity gene segment) (Early *et al.*, 1980; Sakano *et al.*, 1980; Schilling *et al.*, 1980). The formation of a complete V_H gene requires the joining of one V_H, one D_H and one J_H gene segment. Joining of D_H and J_H segments occurs first, followed by fusion of a V_H to the rearranged D_H–J_H gene segments (Alt et al., 1984). At least 12 D_H gene segments are spread over 80 kb of DNA upstream of the four J_H gene segments (Wood and Tonegawa, 1983). The D_H gene segments have the same 5'-to-3' orientation as the J_H and C_H gene segments. The D_H gene segment encodes part of the third hypervariable region and therefore plays an important role in antigen binding.

The orientation and the total number of V_H gene segments, located upstream of the D_H gene segments, are unknown. Current estimates count about 100 V_H gene segments which can be grouped into discrete gene families. Seven such families, each comprising between one and 60 members, have been identified so far and for some of them it has been shown that members of individual families are clustered and are not interspersed with members of other families (Brodeur and Riblet, 1984; Brodeur et al., 1984). Four families have been ordered with respect to their arrangement on chromosome 12: centromere–V_H J558–V_H S107–V_H Q52–V_H7183–D_H–J_H–C_μ. Interestingly, members of the V_H7183 gene family, which are closest to the constant-region genes, appear to rearrange first during B-cell ontogeny (Yancopoulos et al., 1984; Perlmutter et al., 1985).

There are therefore fundamental similarities as well as differences in the mode of DNA rearrangements which occur in light and heavy chain genes. A light chain gene is formed by a process called V–J joining which will juxtapose one of many V gene segments with one of a small number of J gene segments. A heavy chain gene, in contrast, is formed by a process called V–D–J joining which closely resembles the V–J joining of light chain genes but involves a third gene segment, obviously to increase diversity in an important region of the antibody molecule (see below).

4   Heavy chain class switching

Heavy chain genes can undergo class switching which will place any rearranged variable-region gene in front of any one of the seven constant-region genes located downstream of C_μ (Davis et al., 1980; Sakano et al., 1980; Honjo, 1983). After V–D–J joining, at the stage of a pre-B cell, an mRNA molecule is produced which consists of the V region sequence linked to the C_μ constant-region sequence. This allows synthesis of a μ heavy chain. At later stages of B-cell differentiation, however, heavy chain switching will occur which replaces the $C\mu$ gene with any one of the other constant-region genes. For example, the rearranged V region sequence can be fused on the $C_{\gamma1}$ gene, leading to the synthesis of a $\gamma 1$ chain containing the same V region sequence previously associated with the μ chain. Heavy chain switching results in deletion of the intermediate DNA, in this example of the C_μ, C_δ and $C_{\gamma 3}$ genes.

5   *Immunoglobulin gene rearrangements during B-cell differentiation*

Differentiation of B-cell precursors into antibody-producing cells can be divided into phases, antigen independent and antigen dependent (Cooper *et al.*, 1984). Antigen-independent events occur early in B-cell differentiation. Rearrangement of variable-region gene segments occurs first for the heavy chain genes, resulting in the appearance of μ heavy chains in the cytoplasm of pre-B cells. These pre-B cells will subsequently rearrange their light chain genes and express an IgM molecule on the cell surface. The cells are now called B cells. Each B cell will express only one heavy and one light chain gene (allelic and isotypic exclusion) and therefore has committed itself to the detection of a certain antigenic determinant. Contact with this determinant together with stimulation by helper T cells will lead to clonal expansion and further differentiation into antibody-secreting B cells, called plasma cells. Activation of B cells by antigen and T cells will also initiate the formation of memory B cells, which can mount a more vigorous secondary immune response, and the switch from IgM to other classes of heavy chain. Thus, after antigen activation the effector functions of immunoglobulin molecules can be changed while the recognition unit is left intact. In addition, however, somatic mutation events (see below) can introduce amino acid changes into the variable regions of heavy and light chain genes to produce antibodies with even higher affinities for the immunizing antigen.

IV   T-cell receptor genes

A   The T-cell receptor complex

Several molecules on the surface of a T cell appear to be associated with the molecular entity that recognizes MHC and foreign antigen (Owen, 1984). The basic receptor unit appears to be a heterodimeric disulphide-linked transmembrane glycoprotein composed of one α chain and one β chain, both about 40 000–50 000 daltons in molecular weight (Fig. 1).

This molecule is currently called the T-cell receptor and basically looks like a two-chain immunoglobulin molecule composed of two variable-region and two constant-region domains. Some evidence exists that the molecule recognizes both MHC and antigen (Hannum *et al.*, 1984).

The receptor molecule is non-covalently associated in the cell membrane with the T3 molecule (primarily characterized in man (van den Elsen *et al.*, 1984)). T3 is composed of three polypeptide chains, with molecular weights ranging from 20 000 to 25 000 daltons. In contrast with the α and β chains of the T-cell receptor, which show sequence variation in their variable regions if T cells of different specificities are compared, the T3 polypeptide chains are monomorphic. They are believed to be important for certain effector functions, perhaps for the transfer of signals to the inside of the cell.

At least two more molecules appear to play a role in the recognition of its target structure by a T lymphocyte. These are the murine L3T4 and Lyt-2,3 molecules, which are present on T cells that recognize respectively MHC class II and MHC class I molecules as restriction elements. Like the T3 molecule, the L3T4 and Lyt-2,3 molecules are monomorphic. They are believed to enhance the overall avidity of interactions between T cells and target cells by binding to non-polymorphic determinants on the MHC molecule. The binding of L3T4 or Lyt-2,3 molecules to class II or class I molecules respectively appears to be especially important when low affinity binding is exerted through the T-cell receptor (Robertson, 1985).

B   Organization and rearrangement of T-cell receptor α and β chain genes

1 β chain genes

Two constant-region genes, $C_{\beta 1}$ and $C_{\beta 2}$, are located next to each other on chromosome 6 in the mouse, about 6 kb apart (Fig. 5). Both $C_{\beta 1}$ and $C_{\beta 2}$ are flanked by a cluster of seven J_{β} gene segments and one D_{β} gene segment on their 5' side (Gascoigne *et al.*, 1984; Kavaler *et al.*, 1984; Siu *et al.*, 1984). In each cluster, only six of the seven J_{β} genes appear to be functional. The total number of V_{β} gene segments, thought to be located upstream of the C_{β} genes with one exception, is not known but is estimated to be between 16 and 21 (Patten *et al.*, 1984; Barth *et al.*, 1985; Behlke *et al.*, 1985). The $V_{\beta 14}$ gene segment has been found about 10 kb downstream of the $C_{\beta 2}$ gene in an inverted transcriptional orientation (Malissen *et al.*, 1986). This gene segment can be used to produce a functional β chain gene by chromosomal inversion.

Chromosome

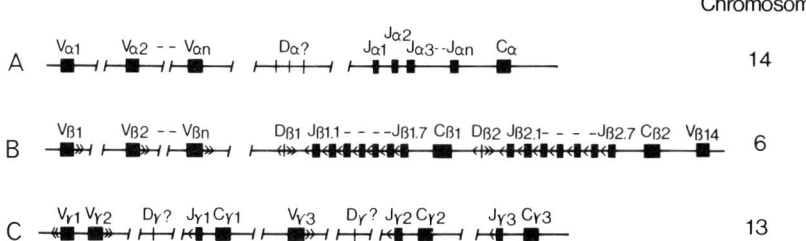

Figure 5 Organization of the three T-cell receptor gene families α(A), β(B) and γ(C), located on mouse chromosomes 14, 6 and 13 respectively. For explanation see the text and the legend to Fig. 3. The order and the orientation of unlinked gene segments are not known. (Not drawn to scale).

The two C_β genes code for two distinct but very similar constant regions of T-cell receptor β chains. They differ in only four (Gascoigne *et al.*, 1984) or five (Malissen *et al.*, 1984) amino acid residues, all located in the carboxy-terminal portion of the β chain. The 3′-untranslated regions of the two C_β genes, however, are very different from one another, suggesting that correction mechanisms (perhaps gene-conversion-like events) and selective forces keep the two coding sequences alike.

The split of the T-cell receptor β chain gene into four segments, V_β, D_β, J_β and C_β, is very similar to the division of immunoglobulin heavy chain genes into four segments, V_H, D_H, J_H and C_H. Rearrangements of β chain gene segments in T cells to produce a functional β chain gene also result in V–D–J joining, although direct V–J joins appear to occur as well (see below).

As for immunoglobulin heavy chain genes, the joining of T-cell receptor D_β and J_β gene segments precedes V_β–D_β–J_β joining. Indeed, the first β chain transcripts observed during thymic ontogeny are about 1 kb in length and are presumably derived from β chain genes which have joined D_β and J_β gene segments only. The small transcripts are subsequently replaced by full-length V_β–D_β–C_β transcripts about 1.3 kb in length (Born *et al.*, 1985; Kronenberg *et al.*, 1985; Snodgrass *et al.*, 1985b).

Functional β chain gene rearrangements occur in helper and cytotoxic T cells. For suppressor T cells, however, conflicting results have been obtained. Several mouse suppressor hybridomas analysed have lost the β chain genes on both homologues of chromosome 6 derived from the functional T cell, while they retain the β chain genes from the thymoma used in the cell fusion (Hedrick *et al.*, 1985; Kronenberg *et al.*, 1985). The deletion of the β chain genes might be due to selective chromosome losses. However, a cloned mouse suppressor T-cell line has been described which apparently contains a functional β chain gene (Mori *et al.*, 1985; Adorini *et al.*, 1986). The same seems to be true for several human suppressor T-cell lines analysed (Royer *et al.*, 1984; Toyonaga *et al.*, 1984). This indicates that at least some suppressor T cells use the same β genes as helper and cytotoxic T cells to code for part of their antigen–MHC receptor.

2 α *chain genes*

In comparison with the β chain genes much less is currently known about the organization and rearrangement of α chain genes. Most of what we know is based on the analysis of α chain cDNA clones which suggest that the α chain gene is composed of V_α, J_α and C_α gene segments and perhaps D_α gene segments (Chien *et al.*, 1984a; Saito *et al.*, 1984b) (Fig. 5). Using V region sequences as probes, DNA rearrangements can clearly be detected in T-cell DNA in contrast with non-T-cell DNA. The mouse contains a single α chain constant-region gene, located on chromosome 14 and closely linked to the gene coding for purine nucleoside phosphorylase (Dembić *et al.*, 1985). This constant-region gene is used in helper and cytotoxic T cells to code for the T-cell receptor α chain (Chien *et al.*, 1984a; Saito *et al.*, 1984b). Using the constant-region gene as a probe, rearrangements are usually not detected in T cells. This is explained by the fact that a large number of J gene segments (more than 18) are spread over at least 63 kb of DNA in front of the constant-region gene (Winoto *et al.*, 1985).

C The T-cell-receptor-related γ chain gene

A third gene family (γ) has been identified which undergoes DNA rearrangements in T cells and is homologous to the genes encoding the α and β chains of the T-cell receptor (Saito *et al.*, 1984a). The γ gene family is composed of at least three V gene segments and three constant-region genes, each of which is flanked on its 5' side by a single J gene segment (Hayday *et al.*, 1985) (Fig. 5). In the mouse the γ genes are located on chromosome 13 (Kranz *et al.*, 1985a). Apparently only one C_γ, one J_γ and one V_γ gene segment are used to code for the γ chain, so far identified at the cDNA but not yet at the protein level. The second constant-region gene appears to be non-functional because of an inappropriate splice site and the third appears to rearrange to a sequence that bears no homology to the three V_γ gene segments identified.

The γ chain gene, although rearranged in both cytotoxic and helper T cells, appears to be transcribed almost exclusively in cytotoxic T cells although at a lower level than α and β chain genes (Kranz *et al.*, 1985b). If this finding indicates a role for the γ chain in the recognition of MHC class I molecules then one might expect to find in helper T cells another chain that is involved in the recognition of class II molecules. Another clue to the possible function of the γ chain might come from the observation that an inverse correlation exists between the transcription of α and γ chain genes during T-cell ontogeny (Raulet *et al.*, 1985; Snodgrass *et al.*, 1985a). Transcription of the γ chain gene occurs first (on or before day 14) but is gradually replaced by α chain gene transcription (starting at about day 17) in the developing thymus. Interestingly, β chain gene transcription starts at about day 15 and the level of transcription does not change much during ontogeny. Perhaps early embryonic thymocytes express a putative γ chain at the cell surface (with or without a second chain) whereas the surface expression of α and β chains occurs at a later stage of T-cell differentiation. If this is true, then recognition of certain molecules on thymic epithelial cells (e.g. MHC molecules) might tell the T cell expressing the γ chain that it is in the thymus and generate a signal for the cell to rearrange its α and β genes. Thus the putative γ chain might be involved in MHC–antigen recognition in immature and mature T cells (Pernis and Axel, 1985).

V   The immunoglobulin superfamily

A   Sequence relationships

Sequence comparisons have shown that the members of the MHC class I and class II, the immunoglobulin and the T-cell receptor gene families are all related (Hood *et al.*, 1985). In fact several other genes, those encoding Thy1, the poly-Ig receptor, T4 (the human homologue of L3T4), T8 (the human homologue of Lyt-2,3) and OX-2, are also known to be members of the immunoglobulin superfamily and one can expect that more members will be found in the future (Williams, 1985).

A priori, homologies observed at the protein level might be due to convergent or divergent evolution of genes: convergent evolution occurs when two independent genes drift towards a common structure because they serve similar functions, whereas divergent evolution describes the further development of genes derived from a common ancestor. Several reasons, some of them discussed below, indicate that the members of the immunoglobulin superfamily were generated by divergent evolution.

B   The immunoglobulin homology unit

As mentioned above, immunoglobulin molecules are composed of various domains serving different functions. These domains are related to one another by sequence homology. Furthermore, they are encoded by distinct exons, indicating that they too were generated during evolution by duplication events. A typical immunoglobulin domain, called the immunoglobulin homology unit, is composed of about 110 amino acids, contains a centrally placed disulphide bridge spanning about 65 amino acid residues and is encoded by a single exon. Introns, separating individual exons from one another, typically are located so that they split the three nucleotides of a codon between the first and the second position. MHC class I, MHC class II and T-cell receptor molecules show clear evidence that they too are composed of immuno-globulin homology units (Hood *et al.*, 1983, 1985; Davis, 1985). In particular the $\alpha3$ domain of class I molecules, the $\alpha2$ and $\beta2$ domains of class II molecules, the variable-region and constant-region domains of the T-cell receptor and β_2-microglobulin show significant amino acid sequence homologies to immunoglobulin domains and fulfil the criteria listed above for the immunoglobulin homology unit.

With respect to the amino-terminal domains of MHC molecules ($\alpha1$ and $\alpha2$ of class I and $\alpha1$ and $\beta1$ of class II), the possibility that they are of independent origin and have been combined with exons encoding the immunoglobulin homology unit cannot be excluded. Exons can be moved from one gene to another. The epidermal growth factor receptor and low density lipoprotein receptor genes, for example, have very homologous exons which they combine with unrelated exons (Yama-moto *et al.*, 1984). In contrast, the similarity in size, the position of the introns and the fact that some of the exons encode a centrally placed disulphide bridge make it more likely that the exons encoding the amino-terminal MHC domains have also been derived from the primor-dial gene encoding the immunoglobulin homology unit.

C   Rearranging immunoglobulin and T-cell receptor gene segments

Of the three gene families discussed in this review, only the immunoglobulin and T-cell receptor gene families undergo rearrangements in cells which express them. Rearrangements have not been found for MHC class I and class II genes.

The precise mechanism used to rearrange immunoglobulin and T-cell receptor V, D and J gene segments remains elusive, despite extensive efforts to clarify it. Several different mechanisms have been discussed. For instance, gene segments could be rearranged by copying and insertion, by excision and insertion, by gene inversion, by deletion of intervening DNA and joining of the segments and by sister chromatid exchange. Evidence for and against the deletion model (Seidman *et al.*, 1980; Steinmetz *et al.*, 1980), for and against the inversion model (Wood and Tonegawa, 1983; Lewis *et al.*, 1984; Reynaud *et al.*, 1985; Malissen *et al.*, 1986) and for and against the sister chromatid exchange model (Seidman *et al.*, 1980; Höchtl *et al.*, 1982; van Ness *et al.*, 1982) has been obtained. Perhaps several of these mechanisms can be used (Kronenberg *et al.*, 1985; Malissen *et al.*, 1986). To escape the present dilemma one needs more information on the enzymatic machinery and perhaps a cell-free assay system (Desiderio and Baltimore, 1984).

Immunoglobulin and T-cell receptor genes display highly conserved sequences at the 3′ end of V, at the 5′ end of J and at the 5′ and 3′ ends of D gene segments (Max *et al.*, 1979; Sakano *et al.*, 1979a; Malissen *et al.*, 1984; Davis, 1985; Hayday *et al.*, 1985). The basic structure of the conserved sequence consists of a conserved heptamer and a conserved nonamer separated by a spacer of either 11–12 or 22–23 nucleotides which is not conserved in sequence. It has been noted that the length of the spacer segment corresponds to either one or two turns of the DNA helix (Early *et al.*, 1980; Sakano *et al.*, 1980). Joining apparently occurs between gene segments displaying a one- and a two-turn spacer (Figs. 3, 4 and 5). So far no exception has been found to the fusion of gene segments according to the one-turn and two-turn spacer rule.

Immunoglobulin and T-cell receptor gene segments contain heptamer and nonamer sequences that are almost identical, in agreement with their proposed common evolutionary origin. The conserved nature of these sequences furthermore argues that they serve an important function, presumably in the rearrangement process itself. The joining of V, D and J gene segments might be accomplished by two highly specific but distinct enzymes, expressed at an early stage of lymphocyte differentiation, which bind and cleave specifically at the conserved sequence with either a one-turn or a two-turn spacer. In this model, the two distinct enzymes would then associate and form a heterodimer (in agreement with the one-turn and two-turn spacer rule) and subsequently fuse the gene segments (Early *et al.*, 1980; Sakano *et al.*, 1980).

If the same, or closely related, enzymatic machinery is used to rearrange immunoglobulin and T-cell receptor gene segments one might expect to find occasionally rearranged immunoglobulin genes in T cells and rearranged T-cell receptor genes in B cells. Indeed, joining of immunoglobulin heavy chain D_H and J_H gene segments has been found at a high frequency in T cells (Kurosawa *et al.*, 1981), and rearrangements of T-cell receptor α, β and γ chain genes have been observed in a B myeloma cell (Traunecker *et al.*, 1986).

The switching of immunoglobulin heavy chain constant-region genes does not involve sequences similar to those which have been found flanking V, J and D gene segments. Instead, simple tandemly repeated sequences have been found in the so-called switch regions upstream of each heavy chain constant-region gene except for the C_δ gene (Honjo, 1983). It has been proposed that switching occurs by unequal sister chromatid exchange but conflicting evidence has recently been found in a transformed line which switched *in vitro* (Wabl *et al.*, 1985b).

No evidence exists for switching between the T-cell receptor $C_{\beta1}$ and $C_{\beta2}$ genes (Kronenberg *et al.*, 1985). Also, switching of immunoglobulin C_λ gene segments, which show a similar organization to the T-cell receptor C_β gene segments, has never been observed.

D   Chromosomal locations and translocations

In Table 1 the chromosomal locations of the genes that are members of the immunoglobulin superfamily are listed for mouse and man. Some members of the superfamily map to the same chromosome, supporting the idea that duplication events have generated the different genes. For instance, in mouse and man the immunoglobulin κ chain gene is tightly linked to the gene(s) coding for the T-cell differentiation antigen Lyt-2,3 (mouse) or T8 (man) (Gibson *et al.*, 1978; Sukhatme *et al.*, 1985). In mice these two genes are also linked to the T-cell receptor β chain gene (Caccia *et al.*, 1984; Lee *et al.*, 1984; Epstein *et al.*, 1985). Furthermore, in men the immunoglobulin heavy chain genes and the T-cell receptor α chain genes are both located on chromosome 14 and the T-cell receptor β and γ chain genes are both located on chromosome 7, although at opposite ends of the chromosomes (Barker *et al.*, 1985; Caccia *et al.*, 1985; Isobe *et al.*, 1985; LeBeau *et al.*, 1985; Morton *et al.*, 1985; Murre *et al.*, 1985; Rabbitts *et al.*, 1985).

It should be pointed out that evidence for sharing of gene segments between different gene families has so far not been found, even among genes located on the same chromosome. Lack of sufficient sequence

Table 1 Chromosomal location of selected genes of the immunoglobulin superfamily in mouse and man.

Gene	Mouse	Man
Immunoglobulin:		
λ	16	22q11
κ	6	2p11
Heavy	12	14q32
T-cell receptor:		
α	14	14q11–q12
β	6	7q32–q36
γ	13	7p15
MHC	17	6p21
β_2-microglobulin	2	15q21–q22
Lyt-2, 3 (T8)	6	2p12
Thy1	9	11q23

homology explains why immunoglobulin V gene segments cannot be used to identify T-cell receptor sequences in hybridization experiments (Kronenberg *et al.*, 1983). Thus there is so far no molecular basis for the hypothesis that T-cell receptor and immunoglobulin molecules share certain antigenic determinants (called idiotypes) because they use V region genes of the same gene pool (Jensenius and Williams, 1982). It is possible that similar idiotypes are formed by V region sequences which are identical at certain critical residues but are otherwise quite different from one another (Davis, 1985).

At least some of the T-cell receptor genes may undergo chromosome translocations in T-cell leukaemias similar to those found for immuno-globulin genes in B-cell myelomas and lymphomas (Klein, 1983). Such translocations apparently lead to the activation of certain cellular oncogenes (see Hall, this volume). For instance, in ataxia telangiectasia patients T-cell leukaemias are often found with chromosomal break-points at 7p14 (T-cell receptor γ chain gene), 7q35 (T-cell receptor β chain gene), 14q12 (T-cell receptor α chain gene) and 14q32 (immunoglo-bulin heavy chain gene) (Fiorilli *et al.*, 1985). Similarly, chromosome breakages and translocations or inversions are found close to the location of the T-cell receptor α chain gene in human T-cell acute and chronic lymphocytic leukaemias and adult T-cell leukaemias (Collins *et al.*, 1985; Croce *et al.*, 1985; Jones *et al.*, 1985; Rabbitts *et al.*, 1985).

VI   Generation of diversity

Two-thirds of the 3×10^8 lymphocytes in a mouse are T cells and one-third are B cells. If each B cell synthesizes a different antibody molecule, a maximum of 10^8 different antibodies could be made in every mouse. The total number of different antibody molecules present in a mouse is not known but is estimated to be of the order of several million (Jerne, 1971). As indicated above, it is now known that the diversity of antibody molecules is a result of various somatic diversification and mutation events which operate on several hundred germline gene segments.

Over the last eight years, since the first isolation of an immunoglobulin variable-region gene (Tonegawa *et al.*, 1977), we have learned a lot about the various mechanisms that contribute to the diversity of antibody molecules. It appears that very similar strategies are being used to generate large B-cell and T-cell repertoires. In contrast with antibody and T-cell receptor genes, which use various somatic diversification mechanisms to expand a limited amount of genetic information, the MHC class I and class II genes are not diversified somatically. However, when different individuals are compared, an impressive variability is observed between MHC alleles. Thus selective pressures have operated on MHC genes to create diversity among the different members of a species whereas they have forced antibody and T-cell receptor genes to diversify in each individual.

It is reasonable that a system concerned with the detection of a limitless number of foreign antigens must develop ways which allow it to synthesize a correspondingly limitless number of different receptor molecules. The variability that is observed for MHC molecules, which in an unknown way participate in the recognition process, is more difficult to understand. The interpretation currently favoured assumes that there are blind spots in the T-cell repertoire resulting from tolerization to self-determinants (Klein, 1984). Since self-tolerization is also MHC restricted (Schwartz, 1984a), individuals expressing different MHC alleles will have different blind spots in their T-cell receptor repertoires. Thus selective pressure might have generated MHC polymorphism to ensure survival of the species when foreign invaders are encountered to which certain individuals are non-responsive.

A   Major histocompatibility complex alleles

Only some of the multiple genes located in the MHC are highly polymorphic and show extensive variation between different alleles whereas others, although perhaps polymorphic, are more conserved (Klein and Figueroa, 1981; Steinmetz and Hood, 1983; Mengle-Gaw and McDevitt, 1986; Steinmetz, 1986). For instance, different K allomorphs have been shown to differ by as much as 16% of their amino acid residues, and the highly variable class II molecules A_α, A_β and E_β differ by as much as 9% between different allomorphs. In contrast, two different E_α allomorphs which have been sequenced differ in only two residues out of a total of 234 and it appears that the Qa and Tla region class I genes, which do not encode T-cell restriction elements, are also conserved (Yokoyama *et al.*, 1983; Mellor *et al.*, 1984; Lalanne *et al.*, 1985). Interestingly, the variability observed between different class I and class II allomorphs occurs preferentially in the amino-terminal domains ($\alpha 1$ and $\alpha 2$ of class I, $\alpha 1$ and $\beta 1$ of class II) and it has been shown by *in vitro* mutagenesis experiments that the specificity exerted by T cells to MHC determinants maps to these domains. In some cases hypervariable regions could be identified in the amino-terminal domains, thus identifying the residues of the molecule most probably involved in binding specificity. It is not known whether these residues make contact with the antigen or the T-cell receptor or with both.

At least three different mechanisms appear to generate the variability in MHC alleles during evolution: gene duplication events together with genetic drift, frequent recombination events at recombinational hot spots and gene conversion events leading to the change of several nucleotides in a small stretch of DNA. Natural selection will favour useful mutants generated by these mechanisms and establish them at reasonable frequencies in the species.

1 *Gene duplication*

Perhaps the most convincing evidence that gene duplication together with genetic drift contribute to the generation of biologically important new MHC alleles comes from the class I genes encoded in the D region. Different mouse strains encode different numbers of class I molecules in this region. For instance, two class I molecules, H-2D^d and H-2L^d, are encoded in the D region of the BALB/c mouse whereas the D region of the C57BL/10 mouse codes for only one molecule, termed H-2D^b (Hood *et al.*, 1983). The genes encoding the three class I molecules have all been cloned and sequence and restriction map comparisons show a close relationship between all three (Taylor Sher *et al.*, 1985; Steinmetz *et al.*, unpublished results), indicating that they have been derived from a common ancestor. The presence of two D-like genes in the BALB/c mouse could be the result of a gene duplication event caused by unequal crossing-over. It is also possible that a gene contraction event reduced the number of D-region class I genes in the C57BL/10 mouse. In fact, such a gene contraction event has apparently generated the H-2D/L fusion gene in the B10.D2(H-2^{dm1}) mutant strain (Burnside *et al.*, 1984; Sun *et al.*, 1985). Whatever the mechanism, extensive sequence homologies between H-2D^d and H-2L^d in 5′ and 3′ flanking sequences indicate that the origin of the two genes was a fairly recent event. Nevertheless, genetic drift, either before or after the duplication event, has allowed the two genes to become functionally distinguishable. For instance, the H-2L^d molecule (but not H-2D^d) is used as a restriction element in the recognition of vesicular stomatitis virus by cytotoxic T cells (Ciavarra and Forman, 1982). In contrast, influenza virus is recognized by cytotoxic T lymphocytes in the context of either the H-2D^d or the H-2L^d molecule (Reiss *et al.*, 1983).

2   Recombination at hot spots

Three hot spots for recombination have recently been identified in the MHC, one located in the middle of the E$_\beta$ gene and two others between the K and A$_{\beta2}$ genes (Shiroishi *et al.*, 1982; Steinmetz *et al.*, 1982a, 1986). It is likely that additional hot spots exist, most probably between the S and D regions of the MHC, because a large number of recombination events have been localized there (Klein *et al.*, 1983a). Interestingly, at least one of the recombinational hot spots that have been mapped (the one in the E$_\beta$ gene) coincides with a boundary separating a tract of DNA with high nucleotide variability from a tract with low nucleotide variability (Steinmetz *et al.*, 1984). Surprisingly, the variability is observed in coding and non-coding sequences. Variable tracts contain the genes encoding the highly polymorphic and variable MHC allomorphs (K, A$_\alpha$, A$_\beta$ and E$_\beta$), while the conserved tracts contain genes that do not show much allelic variation (E$_\alpha$, C4 and Slp).

A possible explanation for the occurrence of variable tracts of DNA in the MHC is that during the process of speciation these sequences—because they carried useful alleles—were extracted from all the different haplotypes available and placed into the MHC of the newly arising species by the process of recombination. The finding of recombinational hot spots at boundaries between variable and conserved tracts supports such a model (Arden and Klein, 1982) of trans-specific evolution of MHC alleles. It is clear that recombination (unless it occurs inside a gene) does not really generate new alleles but it may preserve a battery of useful alleles that have been generated over a long period of evolutionary time.

3   Gene conversion

It appears that gene-conversion-like events, resulting in the replacement in an acceptor gene of a short stretch of nucleotides (perhaps up to 50 bp) by sequences derived from a donor gene, is a major mechanism to generate diversity in certain class I and class II genes (Hood *et al.*, 1983; Steinmetz and Hood, 1983; Steinmetz, 1986). The best evidence stems from the characterization of 14 mutants of the H-2K^b gene and of one mutant of the A$_\beta^b$ gene. For several of the mutants, multiple amino acid changes were identified which occurred in clusters and, in some cases, repeatedly in independently isolated mutants. Two of the H-2K^b mutant genes, H-2K^{bm1} and H-2K^{bm6}, and the A$_\beta^b$ mutant gene, called A$_\beta^{bm12}$, have now been sequenced and compared with the parental genes. In all three cases clustered nucleotide changes were observed: K^{bm1}, seven changes in a stretch of 13 nucleotides (Schulze *et al.*, 1983; Weiss *et al.*, 1983); K^{bm6}, two changes in 15 nucleotides (Nathenson *et al.*, 1985); A$_\beta^{bm12}$, three changes in 14 nucleotides (McIntyre and Seidman, 1984). Moreover, potential donor genes have been identified for all three mutant genes: the Q10 gene, located in the Qa region, could have been used as the donor gene in a gene conversion event generating the H-2K^{bm1} mutant gene (Mellor *et al.*, 1983), another Qa region gene could have been used as the donor gene to generate the H-2K^{bm6} mutant (Nathenson *et al.*, 1985) and the E$_\beta^b$ gene could have been the donor gene to derive the A$_\beta^{bm12}$ mutant (Denaro *et al.*, 1984; Mengle-Gaw *et al.*, 1984; Widera and Flavell, 1984).

B   Immunoglobulin and T-cell receptor genes

So far I have described mechanisms which operate during evolution and create variability in MHC class I and class II genes. It appears likely that similar mechanisms operate on other multigene families for which polymorphism and variability of the encoded gene products are of selective advantage.

The immunoglobulin and T-cell receptor genes, however, use additional mechanisms, distinct from the ones discussed above, to generate diversity throughout the lifetime of an individual. Both multigene families use essentially the same strategies to diversify their genes, which is not surprising considering their evolutionary relationship. Table 2 lists the various mechanisms contributing to the immense diversity of immunoglobulin and T-cell receptor molecules. Operationally one can distinguish between diversity that is already present in germline DNA in the form of multiple gene segments that encode variable regions and diversity that is generated somatically through combinatorial joining and mutation of the germline-encoded gene segments (Tonegawa, 1983; Barth *et al.*, 1985; Behlke *et al.*, 1985; Davis, 1985; Hood *et al.*, 1985).

The somatic diversification mechanisms will also generate non-functional immunoglobulin and T-cell receptor genes at a relatively high rate. This appears to be the price that the immune system has to pay to generate its immense repertoire of receptor molecules.

Table 2   Generation of diversity in mouse immunoglobulin and T-cell receptor genes.

	Immunoglobulin genes	T-cell receptor genes
Multiplicity of germline encoded V region gene segments	Light chain genes 90–300 Vκ 4 J$_\kappa$ 2 V$_\lambda$ 3 J$_\lambda$	α chain genes > 40 Vα ? Dα > 18 J$_\alpha$
	Heavy chain genes 100–200 V$_H$ 12 D$_H$ 4 J$_H$	β chain genes 16–21 V$_\beta$ 2 D$_\beta$ 12 J$_\beta$
		γ chain genes 1 V$_\gamma$? D$_\gamma$ 1 J$_\gamma$
Somatic diversification mechanisms	Combinatorial joining	Combinatorial joining
	Junctional flexibility	Junctional flexibility
	N region diversity	N region diversity
	—	Translational flexibility
	Combinatorial association	(Likely to occur)
	Somatic hypermutation	(Unlikely to occur)

1 *Diversity encoded in germline DNA*

As discussed above, several hundred gene segments coding for immuno-globulin molecules and multiple gene segments coding for T-cell receptor molecules have been identified. The fact that in both families the 3′ end of the V region is split into two or three gene segments provides a strong argument that variability at this position is of major functional importance. This is supported by X-ray crystallographic analyses of immunoglobulin molecules showing that this portion of heavy and light chain variable regions forms part of the antigen-combining site. Indeed the various somatic mechanisms which increase the diversity of antibody and T-cell receptor molecules also focus on the 3′ end of the variable regions.

2 *Diversity generated through somatic rearrangement of variable-region gene segments*

Four different mechanisms, combinatorial joining, flexible joining, random nucleotide insertion and translational flexibility, amplify the diversity encoded in germline gene segments during somatic rearrange-ment processes.

First, combinatorial joining of V and J or V, D and J gene segments enables the cells to generate a large number of V region genes from a relatively small number of gene segments. For example, 100 V_κ and four J_κ gene segments can be assembled into $100 \times 4 = 400$ different V_κ region genes. In case of the T-cell receptor β chain gene segments, the number of V region genes that can be synthesized through combinatorial joining of V, D and J gene segments can be further enlarged by fusing V and J directly (Yoshikai *et al.*, 1984; Rubb *et al.*, 1985) and, perhaps, by D–D joining (see above).

Second, flexible joining describes the observation that V–J and V–D–J fusion is not precise. Flexibility in the exact point of joining exists, so that coding information from the ends of the gene segments can be skipped and amino acids at the recombination sites can be encoded either by the V, D or J or partially by V and D, V and J or D and J gene segments.

Third, during V–D and D–J joining, nucleotides appear to be inserted randomly at the site of recombination leading to the appearance of new amino acids in a critical region of the antibody (N region diversity). It is assumed that the insertion of nucleotides is due to terminal transferase activity.

Fourth, T-cell receptor D_β gene segments show translational flexibility. The two D_β gene segments identified lack translation termination codons in all three reading frames, in contrast with immunoglobulin D_H gene segments. Thus each D_β gene segment can be translated into three different amino acid sequences. By this means β chain genes may partly compensate for their smaller number of D gene segments (only two appear to exist) compared with immunoglobulin heavy chain genes.

Finally, combinatorial association of light and heavy chains contributes in a major way to antibody diversity because the antigen-combining site is formed by hypervariable residues of heavy and light chain variable regions. The random or almost random association of several thousand distinct heavy chains with several hundred distinct light chains in a combinatorial fashion can generate several million different antibody molecules. It is very likely that T-cell receptor molecules will also show combinatorial association of α and β chains.

3   *Somatic hypermutation of variable-region genes*

Somatic hypermutation of immunoglobulin V region genes contributes significantly to antibody diversity. Although the precise mechanism of somatic hypermutation is not known, several characteristic points have emerged recently.

(i) Somatic hypermutation acts exclusively on rearranged V genes but not on V gene segments that stay in the germline configuration (Gorski *et al.*, 1983).

(ii) Somatic hypermutation can be extensive. Mutations are found throughout the V region gene and extend somewhat into the 5′ and 3′ flanking sequences (Kim *et al.*, 1981).

(iii) Somatic hypermutation occurs before and after heavy chain class switching (McKean *et al.*, 1984).

(iv) Somatic hypermutation occurs at a very high rate. Both *in vivo* and *in vitro* experiments indicate that nucleotide changes accumulate in V region genes at the rate of one base change in 10^3–10^5 nucleotides per cell division. This rate is higher by a factor of about 10^4–10^6 than has been observed in other eukaryotic genes (McKean *et al.*, 1984; Wabl *et al.*, 1985a).

(v) Somatic hypermutation events are found preferentially in the hypervariable regions forming the antigen-combining sites of antibody molecules (Tonegawa, 1983). It is reasonable to assume that the observed clustering is due to selection of B lymphocytes which produce antibodies that show an increased affinity for the immunizing antigen. B cells expressing antibodies with higher affinity will be preferentially stimulated by T cells to divide and differentiate and will therefore dominate the immune response.

Somatic hypermutation therefore appears to be due to a mutational mechanism acting before and after class switching on rearranged immunoglobulin heavy and light chain V region genes in a focal manner. Its physiological role appears to reside in the fine tuning of antibody molecules to render them more reactive to the immunizing antigen (Berek *et al.*, 1985).

At the time of writing, only little is known about whether T-cell receptor genes are somatically hypermutated or not. In two cases where germline V_β gene segments have been compared with their rearranged counterparts no evidence for somatic hypermutation has been obtained (Chien *et al.*, 1984b; Goverman *et al.*, 1985). Although base changes have been found in these comparisons, none of them would lead to amino acid changes in the V_β region and it could not be ruled out that they are due to strain polymorphism or sequencing errors. However, it has been found that T-cell receptor genes accumulate mutations when class II alloreactive T-cell hybridomas are grown in culture (Augustin and Sim, 1984; Sim *et al.*, 1985). Whether this *in vitro* observation reflects a physiologically important mechanism for generating diversity in the T-cell repertoire *in vivo* remains to be seen. In fact, a maturation of the T-cell repertoire by somatic hypermutation as it occurs in B cells might be detrimental to the organism, as it could generate self-reactive clones (Barth *et al.*, 1985; Davis, 1985).

VII   Species comparison

In this review I have mainly confined myself to the immunoglobulin, T-cell receptor and MHC genes of the mouse because we know most about these genes in this species. Among the other species that have been studied are man, hamster and chicken. In the following, I shall briefly summarize a few important differences that have been found in these species.

A   Major histocompatibility complex genes

The organization of MHC genes in the HLA complex on chromosome 6 in man appears to be very similar to that in the mouse (Auffray and Strominger, 1986). Twenty to thirty class I genes, six class II α genes, eight class II β genes and, as in the mouse, the same six genes unrelated to class I and class II are present. Slight variations in the gene numbers do occur in different HLA haplotypes. The major difference in gene organization so far is the absence of class I genes proximal to the class II gene loci. Instead all class I genes appear to be clustered at the distal end of the HLA complex, including the three loci (HLA-A, HLA-B, HLA-C) coding for T-cell restriction elements. Some of the class I and class II genes in man are also highly polymorphic whereas others are not. Polymorphism of class I genes has not been found in the hamster, however, although the class II genes are polymorphic. It has been speculated that hamsters might not need polymorphism of class I molecules because they have very little social interaction. If the reason behind the class I polymorphism is indeed the pressure on the immune system to recognize every virus that might spread through a population (as discussed in section VI.A) then a species like the hamster, in which transmission of viruses will be very limited, would not need polymorphic class I molecules (Darden and Streilein, 1984).

B   Immunoglobulin genes

Compared with the mouse the human immunoglobulin genes show a similar complexity and organization (Honjo, 1983). A major difference is the more abundant usage of λ light chains in human immunoglobulin molecules (40% of all light chains are λ chains compared with 5% in the mouse). It therefore appears reasonable that man contains a large number of V_λ genes. Also there are more than six C_λ genes in man compared with four in the mouse. The number of V_κ genes is almost an order of magnitude lower (20–50 V_κ genes in man compared with 90–300 V_κ genes in mouse). The heavy chain constant-region gene cluster in man contains at least ten C_H genes, the major difference from the mouse being the duplication of a chromosomal region containing γ, ε and α genes (Flanagan and Rabbitts, 1982).

Is it possible to generate a large repertoire of antibody molecules from a few germline gene segments? This might be achieved by making more use of somatic hypermutation mechanisms than of combinatorial joining mechanisms of germline gene segments. A good example is the chicken. In this species, different λ light chains, which constitute the major light chain isotype, appear to be generated almost exclusively by somatic hypermutation mechanisms since most if not all of them use a single V_λ gene segment located about 2 kb upstream of a single J_λ–C_λ gene unit (Reynaud *et al.*, 1985). Although about 10 more V_λ gene segments are present further upstream, they all appear to be non-functional. Thus, in each species the various germline and somatic mechanisms that can contribute to antibody diversity may be used to a different extent without significantly changing the size of the final repertoire.

C   T-cell receptor genes

The organization of the T-cell receptor α and β chain gene segments appears to be very similar in mouse and man (Mak and Yanagi, 1984). In contrast, recent experiments have indicated that the γ chain genes display more diversity in man than in mouse (Lefranc and Rabbitts, 1985). While in the mouse a very simplistic pattern of possible rearrangements is observed, more complex rearrangement patterns are seen in human T cells, indicating the existence of a larger number of V_γ, D_γ or J_γ gene segments. Thus, like the human immunoglobulin λ genes, the human γ chain genes appear to be more variable than those in the mouse.

VIII   Conclusions

The immunoglobulin, T-cell receptor and MHC class I and class II gene families clearly are members of a superfamily that appears to have originated from a primordial gene encoding the immunoglobulin homology unit. Other members of the immunoglobulin superfamily have already been identified and surely more will be found in the future. If the reason for the widespread use of the immunoglobulin homology unit is its capacity to interact with itself and to accommodate many different sequences, thus allowing it to interact with other molecular entities, this unit might also be found in receptor molecules which are not associated with the immune system.

Different strategies have evolved to create diversity in MHC class I and class II genes over evolutionary time spans and in immunoglobulin and T-cell receptor genes during the lifetime of an individual. Again it is reasonable to assume that these mechanisms will also be used in other complex biological systems which display molecular entities with variable and conserved functions.

In due course molecular biologists will provide complete linkage maps for the various members of the immunoglobulin, T-cell receptor and MHC gene families. What might be more difficult to determine are the precise molecular interactions that occur during antigen and MHC recognition by the T-cell receptor. Furthermore, we have no idea about the possible function of the large number of class I genes in the Qa and Tla regions of the MHC. It could well be that these genes exert functions related to a possible primordial function of class I molecules before they were engaged by T cells in the process of antigen recognition. Knowledge of these possible primordial functions of MHC molecules might help us to understand why T cells, in contrast with immunoglobulin molecules, need MHC class I and class II molecules to detect foreign antigens.

IX   Acknowledgements

I thank Doug Fisher, Lee Hood, Bernard Malissen, Peter Robinson, John Rogers, Hans Ronne, Henry Sun and Astar Winoto for communication of unpublished results, Z. Dembić, K. Fischer Lindahl and K. Karjalainen for comments, L. Gisler for preparation of the manuscript and H.P. Stahlberger for artwork. The Basel Institute for Immunology was founded and is supported by F. Hoffmann-La Roche, Limited Company, Basel, Switzerland.

X   References

Adorini, L., Palmieri, G., Sette, A., Appella, E. and Doria, G. (1986). Expression of T cell receptor β chain gene products by a mouse monoclonal antigen-specific suppressor T cell line. *Curr. Top. Microbiol. Immunol.*, in press.

Alt, F. W., Yancopoulos, G. D., Blackwell, T. K., Wood, C., Thomas, E., Boss, M., Coffmann, R., Rosenberg, N., Tonegawa, S. and Baltimore, D. (1984). Ordered rearrangement of immunoglobulin heavy chain variable region segments. *EMBO J.* **3**, 1209–1219.

Amor, M., Tosi, M., Duponchel, C., Steinmetz, M. and Meo, T. (1985). Liver mRNA probes disclose two cytochrome P-450 genes duplicated in tandem with the complement C4 loci of the mouse H-2 S region. *Proc. natn. Acad. Sci. USA*, **82**, 4453–4457.

Arden, B. and Klein, J. (1982). Biochemical comparison of major histocompatibility complex molecules from different subspecies of *Mus musculus*: evidence for trans-specific evolution of alleles. *Proc. natn. Acad. Sci. USA* **79**, 2342–2346.

Auffray, C. and Strominger, J. L. (1986). Molecular genetics of the human major histocompatibility complex. *Adv. hum. Genet.*, in press.

Augustin, A. A. and Sim, K. G. (1984). T cell receptors generated via mutations are specific for various major histocompatibility antigens. *Cell* **39**, 5–12.

Barker, P. E., Royer, H.-D., Ruddle, F. H. and Reinherz, E. L. (1985). Regional location of T cell receptor gene Tiα on human chromosome 14. *J. exp. Med.* **162**, 387–392.

Barth, R. K., Kim, B. S., Lan, N. C., Hunkapiller, T., Sobieck, N., Winoto, A., Gershenfeld, H., Okada, C., Hansburg, D., Weissman, I. L. and Hood, L. (1985). The murine T-cell receptor uses a limited repertoire of expressed V$_\beta$ gene segments. *Nature, Lond.* **316**, 517–523.

Behlke, M. A., Spinella, D. G., Chou, H. S., Sha, W., Hart, D. L. and Loh, D. Y. (1985). T-cell receptor β-chain expression: dependence on relatively few variable region genes. *Science, N.Y.* **229**, 566–570.

Berek, C., Griffiths, G. M. and Milstein, C. (1985). Molecular events during maturation of the immune response to oxazolone. *Nature, Lond.* **316**, 412–418.

Bernard, O., Hozumi, N. and Tonegawa, S. (1978). Sequences of mouse immunoglobulin light chain genes before and after somatic changes. *Cell* **15**, 1133–1144.

Blomberg, B., Traunecker, A., Eisen, H. and Tonegawa, S. (1981). Organization of four mouse λ light chain immunoglobulin genes. *Proc. natn. Acad. Sci. USA* **78**, 3765–3769.

Born, W., Yagüe, J., Palmer, E., Kappler, J. and Marrack, P. (1985). Rearrangement of T-cell receptor β-chain genes during T-cell development. *Proc. natn. Acad. Sci. USA* **82**, 2925–2929.

Boyse, E. A. (1984). The biology of Tla. *Cell* **38**, 1–2.

Braunstein, N. and Germain, R. (1985). The murine E$_{\beta2}$ gene: structure, sequence and expression. *In* "Advances in Gene Technology: Molecular Biology of the Immune System" (J. W. Streilein *et al.*, eds) pp. 119–120. Cambridge University Press, London.

Brodeur, P. H. and Riblet, R. (1984). The immunoglobulin heavy chain variable region (Igh-V) locus in the mouse. I. One hundred Igh-V genes comprise seven families of homologous genes. *Eur. J. Immun.* **14**, 922–930.

Brodeur, P. H., Thompson, M. A. and Riblet, R. (1984). The content and organization of mouse IGH-V families. *In* "Regulation of the Immune System" (E. Sercarz, H. Cantor and L. Chess, eds) pp. 445–453. Liss, New York.

Burnside, S. S., Hunt, P., Ozato, K. and Sears, D. W. (1984). A molecular hybrid of the H-2D^d and H-2L^d genes expressed in the dml mutant. *Proc. natn. Acad. Sci. USA* **81**, 5204–5208.

Caccia, N., Kronenberg, M., Saxe, D., Haars, R., Bruns, G. A. P., Goverman, J., Malissen, M., Willard, H., Yoshikai, Y., Simon, M., Hood, L. and Mak, T. W. (1984). The T cell receptor β chain genes are located on chromosome 6 in mice and chromosome 7 in humans. *Cell* **37**, 1091–1099.

Caccia, N., Bruns, G. A., Kirsch, I. R., Hollis, G. F., Bertness, V. and Mak, T. W. (1985). T cell receptor α chain genes are located on chromosome 14 at 14q11–14q12 in humans. *J. exp. Med.* **161**, 1255–1260.

Chaplin, D. D., Woods, D. E., Whitehead, A. S., Goldberger, G., Colten, H. R. and Seidman, J. G. (1983). Molecular map of the murine S region. *Proc. natn. Acad. Sci. USA* **80**, 6947–6951.

Chien, Y. H., Becker, D. M., Lindsten, T., Okamura, M., Cohen, D. I. and Davis, M. M. (1984a). A third type of murine T-cell receptor gene. *Nature, Lond.* **312**, 31–35.

Chien, Y. H., Gascoigne, N. R. J., Kavaler, J., Lee, N. E. and Davis, M. M.

(1984b). Somatic recombination in a murine T-cell receptor gene. *Nature, Lond.* **309**, 322–326.

Ciavarra, R. and Forman, J. (1982). H-2L restricted recognition of viral antigens. *J. exp. Med.* **156**, 778–790.

Collins, M. K. L., Goodfellow, P. N., Spurr, N. K., Solomon, E., Tanigawa, G., Tonegawa, S. and Owen, M. J. (1985). The human T-cell receptor α-chain gene maps to chromosome 14. *Nature, Lond.* **314**, 273–274.

Cooper, M. D., Kearney, J. and Scher, I. (1984). B lymphocytes. *In* "Fundamental Immunology" (W. E. Paul, ed.) pp. 43–55. Raven Press, New York.

Cory, S., Tyler, B. M. and Adams, J. M. (1981). Sets of immunoglobulin V_κ genes homologous to ten cloned V_κ genes: implications for the number of germline V_κ genes. *J. molec. appl. Genet.* **1**, 103–116.

Croce, C. M., Isobe, M., Palumbo, A., Puck, J., Ming, J., Tweardy, D., Erikson, J., Davis, M. and Rovera, G. (1985). Gene for α- chain of human T-cell receptor: location on chromosome 14 region involved in T-cell neoplasms. *Science, N.Y.* **227**, 1044–1047.

Darden, A. G. and Streilein, J. W. (1984). Syrian hamsters express two monomorphic class I major histocompatibility complex molecules. *Immunogenetics* **20**, 603–622.

Davis, M. M. (1985). Molecular genetics of the T cell receptor beta chain. *Ann. Rev. Immun.* **3**, 537–560.

Davis, M. M., Calame, K., Early, P. W., Livant, D. L., Joho, R., Weissman, I. L. and Hood, L. (1980). An immunoglobulin heavy-chain gene is formed by at least two recombinational events. *Nature, Lond.* **283**, 733–739.

Dembić, Z., Bannwarth, W., Taylor, B. A. and Steinmetz, M. (1985). The gene encoding the T-cell receptor α-chain maps close to the Np-2 locus on mouse chromosome 14. *Nature, Lond.* **314**, 271–273.

Denaro, M., Hammerling, U., Rask L. and Peterson, P. A. (1984). The E_β^b gene may have acted as the donor gene in a gene conversion like event generating the A_β^{bm12} mutant. *EMBO J.* **9**, 2029–2032.

Desiderio, S. and Baltimore, D. (1984). Double-stranded cleavage by cell extracts near recombinational signal sequences of immunoglobulin genes. *Nature, Lond.* **308**, 860–862.

Dreyer, W. J. and Bennett, J. C. (1965). The molecular basis of antibody formation: a paradox. *Proc. natn. Acad. Sci. USA* **54**, 864–869.

Early, P., Huang, H., Davis, M., Calame, K. and Hood, L. (1980). An immunoglobulin heavy chain variable region gene is generated from three segments of DNA: V_H, D and J_H. *Cell* **19**, 981–992.

Elliott, B. W., Jr., Eisen, H. N. and Steiner, L. A. (1982). Unusual association of V, J and C regions in a mouse immunoglobulin λ chain. *Nature, Lond.* **299**, 559–561.

van den Elsen, P., Shepley, B. A., Borst, J., Coligan, J. E., Markham, A. F., Orkin, S. and Terhorst, C. (1984). Isolation of cDNA clones encoding the 20K T3 glycoprotein of human T-cell receptor complex. *Nature, Lond.* **312**, 413–418.

Epstein, R., Roehm, N., Marrack, P., Kappler, J., Davis, M., Hedrick, S. and Cohn, M. (1985). Genetic markers of the antigen-specific T cell receptor locus. *J. exp. Med.* **161**, 1219–1224.

Fiorilli, M., Carbonari, M., Crescenzi, M., Russo, G. and Aiuti, F. (1985). T cell receptor genes and ataxia telangiectasia. *Nature, Lond.* **313**, 186.

Fischer Lindahl, K., Hausmann, B. and Chapman, V. M. (1983). A new H-2 linked class I gene whose expression depends on a maternally inherited factor. *Nature, Lond.* **306**, 383–385.

Fisher, D. A., Hunt III, S. W. and Hood, L. E. (1985). Structure of a gene encoding a murine thymus leukemia antigen and organization of Tla genes in the BALB/c mouse. *J. exp. Med.* **162**, 528–545.

Flaherty, L. (1981). Tla-region antigens. *In* "The Role of the Major Histocompatibility Complex in Immunology" (M. E. Dorf, ed) pp. 33–57. Garland STPM, New York.

Flaherty, L., DiBiase, K., Lynes, M. A., Seidman, J. G., Weinberger, O. and Rinchik, E. M. (1985). Characterization of a Q subregion gene in the murine major histocompatibility complex. *Proc. natn. Acad. Sci. USA* **82**, 1503–1507.

Flanagan, J. G. and Rabbitts, T. H. (1982). Arrangement of human immunoglobulin heavy chain constant region genes implies evolutionary duplication of a segment containing γ, ε and α genes. *Nature, Lond.* **300**, 709–713.

Gascoigne, N. R. J., Chien, Y. H., Becker, D. M., Kavaler, J. and Davis, M. M. (1984). Genomic organization and sequence of T-cell receptor β-chain constant- and joining-region genes. *Nature, Lond.* **310**, 387–391.

Gibson, D. M., Taylor, B. A. and Cherry, M. (1978). Evidence for close linkage of a mouse light chain marker with the Ly-2,3 locus. *J. Immun.* **121**, 1585–1590.

Goodenow, R. S., McMillan, M., Nicolson, M., Taylor Sher, B., Eakle, K., Davidson, N. and Hood, L. (1982). Identification of the class I genes of the mouse major histocompatibility complex by DNA-mediated gene transfer. *Nature, Lond.* **300**, 231–237.

Gorer, P. A. (1936). The detection of antigenic differences in mouse erythrocytes by the employment of immune sera. *Br. J. exp. Path.* **17**, 42–50.

Gorski, J., Rollini, P. and Mach, B. (1983). Somatic mutations of immunoglobulin variable genes are restricted to the rearranged V gene. *Science, N.Y.* **220**, 1179–1181.

Goverman, J., Minard, K., Shastri, N., Hunkapiller, T., Hansburg, D., Sercarz, E. and Hood, L. (1985). Rearranged β T-cell receptor genes in a helper T-cell clone specific for lysozyme: no correlation between V_β and MHC restriction. *Cell* **40**, 859–867.

Hämmerling, U., Ronne, H., Widmark, E., Servenius, B., Denaro, M., Rask, L. and Peterson, P. A. (1985). Gene duplications in the TL region of the mouse major histocompatibility complex. *EMBO J.* **4**, 1431–1434.

Hannum, C., Freed, J. H., Tarr, G., Kappler, J. and Marrack, P. (1984). Biochemistry and distribution of the T cell receptor. *Immun. Rev.* **81**, 161–176.

Hayday, A. C., Saito, H., Gillies, S. D., Kraz, D. M., Tanigawa, G., Eisen, H. N. and Tonegawa, S. (1985). Structure, organization and somatic rearrangement of T cell gamma genes. *Cell* **40**, 259–269.

Hedrick, S. M., Germain, R. N., Bevan, M. J., Dorf, M., Engel, I., Fink, P., Gascoigne, N., Heber-Katz, E., Kapp, J., Kaufmann, Y., Kaye, J., Melchers, F., Pierce, C., Schwartz, R. H., Sorensen, C., Taniguchi, M. and Davis, M. M. (1985). Rearrangement and transcription of a T-cell receptor β-chain gene in different T-cell subsets. *Proc. natn. Acad. Sci. USA* **82**, 531–555.

Höchtl, J., Müller, C. R. and Zachau, H. G. (1982). Recombined flanks of the variable and joining segments of immunoglobulin genes. *Proc. natn. Acad. Sci. USA* **79**, 1383–1387.

Honjo, T. (1983). Immunoglobulin genes. *Ann. Rev. Immun.* **1**, 499–528.

Hood, L., Kronenberg, M. and Hunkapiller, T. (1985). T cell antigen receptors and the immunoglobulin supergene family. *Cell* **40**, 225–229.

Hood, L., Steinmetz, M. and Malissen, B. (1983). Genes of the major histocompatibility complex of the mouse. *Ann. Rev. Immun.* **1**, 529–568.

Isobe, M., Erikson, J., Emanuel, B. S., Nowell, P. C. and Croce, C. M. (1985). Location of gene for β subunit of human T-cell receptor at band 7q35, a region prone to rearrangements in T cells. *Science, N.Y.* **228**, 580–582.

Jensenius, J. C. and Williams, A. F. (1982). The T lymphocyte antigen receptor—paradigm lost. *Nature, Lond.* **300**, 583–588.

Jerne, N. K. (1971). The somatic generation of immune recognition. *Eur. J. Immun.* **1**, 1–9.

Jones, C., Morse, H. G., Kao, F. T., Carbone, A. and Palmer, E. (1985). Human T-cell receptor α-chain genes: location on chromosome 14. *Science, N.Y.* **228**, 83–85.

Kabat, E. A., Wu, T. T., Bilofsky, H., Reid-Miller, M. and Perry, H. (1983). "Sequences of Proteins of Immunological Interest." National Institute of Health, Bethesda, MD.

Kavaler, J., Davis, M. M. and Chien, Y. H. (1984). Localization of a T-cell receptor diversity-region element. *Nature, Lond.* **310**, 421–423.

Kim, S., Davis, M., Sinn, E., Patten, P. and Hood, L. (1981). Antibody diversity: somatic hypermutation of rearranged V_H genes. *Cell* **27**, 573–581.

Klein, G. (1983). Specific chromosomal translocations and the genesis of B-cell-derived tumors in mice and men. *Cell* **32**, 311–315.

Klein, J. (1982). "Immunology: the Science of Self–Nonself Discrimination." Wiley, New York.

Klein, J. (1984). What causes immunological nonresponsiveness? *Immun. Rev.* **81**, 177–202.

Klein, J. and Figueroa, F. (1981). Polymorphism of the mouse H-2 loci. *Immun. Rev.* **60**, 23–57.

Klein, J., Figueroa, F. and Klein, D. (1982). H-2 haplotypes, genes, and antigens: second listing. I. Non-H-2 loci on chromosome 17. *Immunogenetics* **16**, 285–317.

Klein, J., Figueroa, F. and David, C. S. (1983a). H-2 haplotypes, genes and antigens: second listing. II. The H-2 complex. *Immunogenetics* **17**, 553–596.

Klein, J., Figueroa, F. and Nagy, Z. A. (1983b). Genetics of the major histocompatibility complex: the final act. *Ann. Rev. Immun.* **1**, 119–142.

Kranz, D. M., Saito, H., Disteche, C. M., Swisshelm, K., Pravtcheva, D., Ruddle, F. H., Eisen, H. N. and Tonegawa, S. (1985a). Chromosomal locations of the murine T-cell receptor alpha-chain gene and the T-cell gamma gene. *Science, N.Y.* **227**, 941–945.

Kranz, D. M., Saito, H., Heller, M., Takagaki, Y., Haas, W., Eisen, H. N. and Tonegawa, S. (1985b). Limited diversity of the rearranged T-cell γ gene. *Nature, Lond.* **313**, 752–755.

Kronenberg, M., Kraig, E. and Hood, L. (1983). Finding the T-cell antigen receptor: past attempts and future promise. *Cell* **34**, 327–329.

Kronenberg, M., Goverman, J., Haars, R., Malissen, M., Kraig, E., Phillips, L., Delovitch, T., Suciu-Foca, N. and Hood, L. (1985). Rearrangement and transcription of the β-chain genes of the T-cell antigen receptor in different types of murine lymphocytes. *Nature, Lond.* **313**, 647–653.

Kurosawa, Y., von Boehmer, H., Haas, W., Sakano, H., Traunecker, A. and Tonegawa, S. (1981). Identification of D segments of immunoglobulin heavy-chain genes and their rearrangement in T lymphocytes. *Nature, Lond.* **290**, 565–570.

Lalanne, J.-L., Transy, C., Guerin, S., Darche, S., Meulien, P. and Kourilsky, P. (1985). Expression of class I genes in the major histocompatibility complex: identification of eight distinct mRNAs in DBA/2 mouse liver. *Cell* **41**, 469–478.

Larhammar, D., Hammerling, U., Denaro, M., Lund, T., Flavell, R. A., Rask, L. and Peterson, P. A. (1983). Structure of the murine immune response I-A$_\beta$ locus: sequence of the I-A$_\beta$ gene and an adjacent β-chain second domain exon. *Cell* **34**, 179–188.

LeBeau, M. M., Diaz, M. O., Rowley, J. D. and Mak, T. W. (1985). Chromosomal localization of the human T cell receptor β chain genes. *Cell* **41**, 335.

Leder, P. (1982). The genetics of antibody diversity. *Sci. Am.* **246** (May issue), 72–83.

Lee, N. E., D'Eustachio, P., Pravtcheva, D., Ruddle, F. H., Hedrick, S. M. and Davis, M. M. (1984). Murine T cell receptor beta chain is encoded on chromosome 6. *J. exp. Med.* **160**, 905–913.

Lefranc, M.-P. and Rabbitts, T. H. (1985). Two tandemly organized human genes encoding the T-cell γ constant-region sequences show multiple rearrangements in different T-cell types. *Nature, Lond.* **316**, 464–466.

Lévi-Strauss, M., Tosi, M., Steinmetz, M., Klein, J. and Meo, T. (1985). Multiple duplications of complement C4 gene correlate with the H-2 controlled testosterone independent expression of its sex-limited isoform. *Proc. natn. Acad. Sci. USA* **82**, 1746–1750.

Lewis, S., Gifford, A. and Baltimore, D. (1984). Joining of V$_\kappa$ to J$_\kappa$ gene segments in a retroviral vector introduced into lymphoid cells. *Nature, Lond.* **308**, 425–428.

Mak, T. W. and Yanagi, Y. (1984). Genes encoding the human T cell antigen receptor. *Immun. Rev.* **81**, 221–233.

Malissen, M., Minard, K., Mjolsness, S., Kronenberg, M., Goverman, J., Hunkapiller, T., Prystowsky, M. B., Yoshikai, Y., Fitch, F., Mak, T. W. and Hood, L. (1984). Mouse T cell antigen receptor: structure and organization of constant and joining gene segments encoding the β polypeptide. *Cell* **37**, 1101–1110.

Malissen, M., McCoy, C., Blanc, D., Trucy, J., Devaux, C., Schmitt-Verhulst, A., Fitch, F., Hood, L. and Malissen, B. (1986). Direct evidence for chromosomal inversion during T-cell receptor β gene rearrangements. *Nature, Lond.* **319**, 28–33.

Maloy, W. L., Coligan, J. E., Barra, Y. and Jay, G. (1984). Detection of a secreted form of the murine H-2 class I antigen with an antibody against its predicted carboxyl terminus. *Proc. natn. Acad. Sci. USA* **81**, 1216–1220.

Max, E. E. (1984). Immunoglobulins: molecular genetics. *In* "Fundamental Immunology" (W. E. Paul, ed.) pp. 167–204. Raven Press, New York.

Max, E. E., Seidman, J. G. and Leder, P. (1979). Sequence of five potential recombination sites encoded close to an immunoglobulin κ constant region gene. *Proc. natn. Acad. Sci. USA* **76**, 3450–3454.

McIntyre, K. R. and Seidman, J. G. (1984). Nucleotide sequence of mutant I-A$_\beta^{bm12}$ gene is evidence for genetic exchange between mouse immune response genes. *Nature, Lond.* **308**, 551–553.

McKean, D., Huppi, K., Bell, M., Staudt, L., Gerhard, W. and Weigert, M. (1984). Generation of antibody diversity in the immune response of BALB/c mice to influenza virus hemagglutinin. *Proc. natn. Acad. Sci. USA* **81**, 3180–3184.

Mellor, A. L., Weiss, E. H., Ramachandran, K. and Flavell, R. A. (1983). A potential donor gene for the bml gene conversion event in the C57BL mouse. *Nature, Lond.* **306**, 792–795.

Mellor, A. L., Weiss, E. H., Kress, M., Jay, G. and Flavell, R. A. (1984). A nonpolymorphic class I gene in the murine major histocompatibility complex. *Cell* **36**, 139–144.

Mellor, A. L., Antoniou, J. and Robinson, P. J. (1985). Structure and expression of class I genes encoding murine Qa-2 antigens. *Proc. natn. Acad. Sci. USA* **82**, 5920–5924.

Mengle-Gaw, L. and McDevitt, H. O. (1985). Genetics and expression of mouse Ia antigens. *Ann. Rev. Immun.* **3**, 367–396.

Mengle-Gaw, L., Conner, S., McDevitt, H. O. and Fathman, C. G. (1984). Gene conversion between murine class II major histocompatibility complex loci. *J. exp. Med.* **160**, 1184–1194.

Michaelson, J., Boyse, E. A., Chorney, M., Flaherty, L., Fleisner, I., Hämmerling, U., Reinisch, C., Rosenson, R. and Shen, F. W. (1983). The biochemical genetics of the Qa–Tla region. *Transpl. Proc.* **15**, 2033–2038.

Miller, J., Bothwell, A. and Storb, U. (1981). Physical linkage of the constant region genes for immunoglobulins λ_I and λ_{III}. *Proc. natn. Acad. Sci. USA* **78**, 3829–3833.

Mori, L., Lecoq, A. F., Robbiati, F., Barbanti, E., Richi, M., Sinigaglia, F., Clementi, F. and Ricciardi-Castagnoli, P. (1985). Rearrangement and expression of the antigen receptor α, β and γ genes in suppressor antigen-specific T cell lines. *EMBO J.* **4**, 2025–2030.

Morton, C. C., Duby, A. D., Eddy, R. L., Shows, T. B. and Seidman, J. G. (1985). Genes for β chain of human T-cell antigen receptor map to regions of chromosomal rearrangement in T cells. *Science, N.Y.* **228**, 582–585.

Murre, C., Waldmann, R. A., Morton, C. C., Bongiovanni, K. F., Waldmann, T. A., Shows, T. B. and Seidman, J. G. (1985). Human γ-chain genes are rearranged in leukaemic T cells and map to the short arm of chromosome 7. *Nature, Lond.* **316**, 549–552.

Nathenson, S. G., Geliebter, J., Geier, S. S., Mashimo, H., Hemmi, S., Kumar, A., McGovern, D., Nakagawa, M., Pfaffenbach, G., Pontarotti, P. and Zeff, R. (1985). The study of H-2 major histocompatibility complex mutants reveals structure–function relationships and mechanisms of generation of diversity and polymorphism. *In* "Advances in Gene Technology: Molecular Biology of the Immune System" (J. W. Streilein *et al.*, eds) pp. 37–40. Cambridge University Press, London.

van Ness, B. G., Coleclough, C., Perry, R. P. and Weigert, M. (1982). DNA between variable and joining gene segments of immunoglobulin κ light chain is frequently retained in cells that rearrange the κ locus. *Proc. natn. Acad. Sci. USA* **79**, 262–266.

Ogata, R. T. and Sepich, D. S. (1984). Genes for murine fourth complement component (C4) and sex-limited protein (Slp) identified by hybridization to C4- and Slp-specific cDNA. *Proc. natn. Acad. Sci. USA* **81**, 4908–4911.

Owen, M. J. (1984). T-cell receptor companions. *Nature, Lond.* **312**, 406.

Parnes, J. R. and Seidman, J. G. (1982). Structure of wild-type and mutant mouse β_2-microglobulin genes. *Cell* **29**, 661–669.

Patten, P., Yokota, T., Rothbard, J., Chien, Y. H., Arai, K. I. and Davis, M. M. (1984) Structure, expression and divergence of T-cell receptor β-chain variable regions. *Nature, Lond.* **312**, 40–46.

Paul, W. E. (ed.) (1984). "Fundamental Immunology." Raven Press, New York.

Perlmutter, R. M., Kearney, J. F., Chang, S. P. and Hood, L. E. (1985). Developmentally controlled expression of immunoglobulin V_H genes. *Science, N.Y.* **227**, 1597–1601.

Pernis, B. and Axel, R. (1985). A one and a half receptor model for MHC-restricted antigen recognition by T lymphocytes. *Cell* **41**, 13–16.

Rabbitts, T. H., Stinson, M. A., Lefranc, M. P., Steinmetz, M., Goodfellow, P. and Schroeder, J. (1985). The chromosomal location of T cell receptor genes and a T cell rearranging gene: possible correlation with specific translocations in human T cell leukaemia. *EMBO J.*, **4**, 1461–1465.

Raulet, D. H., Garman, R. D., Saito, H. and Tonegawa, S. (1985). Developmental regulation of T-cell receptor gene expression. *Nature, Lond.* **314**, 103–107.

Reiss, C. S., Evans, G. A., Margulies, D. H., Seidman, J. G. and Burakoff, S. J. (1983). Allospecific and virus-specific cytolytic T lymphocytes are restricted to the N or Cl domain of H-2 antigens expressed on L cells after DNA-mediated gene transfer. *Proc. natn. Acad. Sci. USA* **80**, 2709–2712.

Reynaud, C. A., Anquez, V., Dahan, A. and Weill, J. C. (1985). A single rearrangement event generates most of the chicken immunoglobulin light chain diversity. *Cell* **40**, 283–291.

Robertson, M. (1985). The preset state of recognition. *Nature, Lond.* **317**, 768–771.

Rogers, J. H. (1985). Family organization of mouse H-2 class I genes. *Immunogenetics* **21**, 343–353.

Royer, H. D., Bensussan, A., Acuto, O. and Reinherz, E. L. (1984). Functional isotypes are not encoded by the constant region genes of the β subunit of the T cell receptor for antigen/major histocompatibility complex. *J. exp. Med.* **160**, 947–952.

Rubb, F., Acha-Orbea, H., Hengartner, H., Zinkernagel, R. and Joho, R. (1985). Identical $V_β$ T cell receptor genes used in alloreactive cytotoxic and antigen plus I-A specific helper T cells. *Nature, Lond.* **315**, 425–427.

Saito, H., Kranz, D. M., Takagaki, Y., Hayday, A. C., Eisen, H. N. and Tonegawa, S. (1984a). Complete primary structure of a heterodimeric T-cell receptor deduced from cDNA sequences. *Nature, Lond.* **309**, 757–762.

Saito, H., Kranz, D. M., Takagaki, Y., Hayday, A. C., Eisen, H. N. and Tonegawa, S. (1984b). A third rearranged and expressed gene in a clone of cytoxic T lymphocytes. *Nature, Lond.* **312**, 36–40.

Sakano, H., Hüppi, K., Heinrich, G. and Tonegawa, S. (1979a). Sequences at the somatic recombination site of immunoglobulin light-chain genes. *Nature, Lond.* **280**, 288–294.

Sakano, H., Rogers, J. H., Hüppi, K., Brack, C., Traunecker, A., Maki, R., Wall, R. and Tonegawa, S. (1979b). Domains and the hinge region of an immunoglobulin heavy chain are encoded in separate DNA segments. *Nature, Lond.* **277**, 627–633.

Sakano, H., Maki, R., Kurosawa, Y., Roeder, W. and Tonegawa, S. (1980). Two types of somatic recombination are necessary for the generation of complete immunoglobulin heavy-chain genes. *Nature, Lond.* **286**, 676–683.

Schilling, J., Clevinger, B., Davie, J. M. and Hood, L. (1980). Amino acid sequence of homogeneous antibodies to dextran and DNA rearrangements in heavy chain V-region gene segments. *Nature, Lond.* **283**, 35–40.

Schulze, D. H., Pease, L. R., Geier, S. S., Reyes, A. A., Sarmiento, L. A., Wallace, R. B. and Nathenson, S. G. (1983). Comparison of the cloned H-2K^bml variant gene with the H-2K^b gene shows a cluster of seven nucleotide differences. *Proc. natn. Acad. Sci. USA* **80**, 2007–2011.

Schwartz, R. (1984a). Induction of tolerance to self in T lymphocytes. *Nature, Lond.* **308**, 690–691.

Schwartz, R. H. (1984b). The role of gene products of the major histocompatibility complex in T cell activation and cellular interactions. *In* "Fundamental Immunology" (W. E. Paul, ed.) pp. 379–438. Raven Press, New York.

Schwartz, R. H. (1985). T-lymphocyte recognition of antigen in association with gene products of the major histocompatibility complex. *Ann. Rev. Immun.* **3**, 237–261.

Seidman, J. G., Nau, M. M., Norman, B., Kwan, S.-P., Scharff, M. and Leder, P. (1980). Immunoglobulin V/J recombination is accompanied by deletion of joining site and variable region segments. *Proc. natn. Acad. Sci. USA* **77**, 6022–6026.

Shimizu, A., Takahashi, N., Yaoita, Y. and Honjo, T. (1982). Organization of the constant region gene family of the mouse immunoglobulin heavy chain. *Cell* **28**, 499–506.

Shiroishi, T., Sagai, T. and Moriwaki, K. (1982). A new wild-derived H-2 haplotype enhancing K-IA recombination. *Nature, Lond.* **300**, 370–372.

Sim, G. K., MacNeil, I. A., Wheat, W. H., Nelson, J. E. and Augustin, A. A. (1985). Mutations in T cell receptor genes. *In* "Advances in Gene Technology: Molecular Biology of the Immune System" (J. W. Streilein *et al.*, eds) pp. 69–72. Cambridge University Press, London.

Siu, G., Kronenberg, M., Strauss, E., Haars, R., Mak, T. W. and Hood, L. (1984). The structure, rearrangement and expression of D_β gene segments of the murine T-cell antigen receptor. *Nature, Lond.* **311**, 344–350.

Snell, G. D. (1981). Studies in histocompatibility. *Science, N.Y.* **213**, 172–178.

Snodgrass, H. R., Dembić, Z., Steinmetz, M. and von Boehmer, H. (1985a). The expression of T cell antigen receptor genes during fetal development within the thymus. *Nature, Lond.* **315**, 232–233.

Snodgrass, H. R., Kisielow, P., Kiefer, M., Steinmetz M. and von Boehmer, H. (1985b). Ontogeny of the T cell antigen receptor within the thymus. *Nature, Lond.* **313**, 592–595.

Steinmetz, M. (1986). Structural and functional studies of mouse class II genes. *In* "Human Class II Histocompatibility Antigens" (S. Ferrone *et al.*, eds), pp. 109–127. Springer, Berlin.

Steinmetz, M. and Hood, L. (1983). Genes of the major histocompatibility complex in mouse and man. *Science, N.Y.* **222**, 727–733.

Steinmetz, M., Altenburger, W. and Zachau, H. G. (1980). A rearranged DNA sequence possibly related to the translocation of immunoglobulin gene segments. *Nucl. Acids Res.* **8**, 1709–1720.

Steinmetz, M., Moore, K. W., Frelinger, J. G., Taylor Sher, B., Shen, F. W., Boyse, E. A. and Hood, L. (1981). A pseudogene homologous to mouse transplantation antigens: transplantation antigens are encoded by eight exons that correlate with protein domains. *Cell* **25**, 683–692.

Steinmetz, M., Minard, K., Horvath, S., McNicholas, J., Frelinger, J., Wake, C.,

Long, E., Mach, B. and Hood, L. (1982a). A molecular map of the immune response region from the major histocompatibility complex of the mouse. *Nature, Lond.* **300**, 35–42.

Steinmetz, M., Winoto, A., Minard, K. and Hood, L. (1982b). Clusters of genes encoding mouse transplantation antigens. *Cell* **28**, 489–498.

Steinmetz, M., Malissen, M., Hood, L., Örn, A., Maki, R. A., Dastoornikoo, G. R., Stephan, D., Gibb, E. and Romaniuk, R. (1984). Tracts of high or low sequence divergence in the mouse major histocompatibility complex. *EMBO J.* **3**, 2995–3003.

Steinmetz, M., Stephan, D. and Fischer Lindahl, K. (1986). Gene organization and recombination hot spots in the murine major histocompatibility complex. *Cell,* in press.

Sukhatme, V. P., Vollmer, A. C., Erikson, J., Isobe, M., Croce, C. and Parnes, J. R. (1985). The gene for the human T cell differentiation antigen Leu-2/T8 is closely linked to the κ light chain locus on chromosome 2. *J. exp. Med.* **161**, 429–434.

Sun, H., Goodenow, R. S. and Hood, L. (1985). Molecular basis of the dm1 mutation in the major histocompatibility complex of the mouse: a D/L hybrid gene. *J. exp. Med.* **162**, 1588–1602.

Taylor Sher, B., Nairn, R., Coligan, J. E. and Hood, L. E. (1985). DNA sequence of the mouse H-2D^d transplantation antigen gene. *Proc. natn. Acad. Sci. USA* **82**, 1175–1179.

Tonegawa, S. (1983). Somatic generation of antibody diversity. *Nature, Lond.* **302**, 575–581.

Tonegawa, S., Brack, C., Hozumi, N. and Schuller, R. (1977). Cloning of an immunoglobulin variable region gene from mouse embryo. *Proc. natn. Acad. Sci. USA* **74**, 3518–3522.

Toyonaga, B., Yanagi, Y., Suciu-Foca, N., Minden, M. and Mak, T. W. (1984). Rearrangement of T-cell receptor gene YT35 in human DNA from thymic leukaemia T-cell lines and fuctional T-cell clones. *Nature, Lond.* **311**, 385–387.

Traunecker, A., Kiefer, M., Dembić, Z., Steinmetz, M. and Karjalainen, K. (1986). Rearrangements of T cell receptor loci can be found in B lymphoid cells. *Eur. J. Immun.,* in press.

Wabl, M., Burrows, P. D., von Gabain, A. and Steinberg, C. (1985a). Hypermutation at the immunoglobulin heavy chain locus in a pre-B-cell line. *Proc. natn. Acad. Sci. USA* **82**, 479–482.

Wabl, M., Meyer, J., Beck-Engeser, G., Tenkhoff, M. and Burrows, P. D. (1985b). Critical test of a sister chromatid exchange model for the immunoglobulin heavy-chain class switch. *Nature, Lond.* **313**, 687–689.

Wake, C. T., Widera, G. and Flavell, R. A. (1985). Organization and expression of the murine MHC. *In* "Advances in Gene Technology: Molecular Biology of the Immune System" (J. W. Streilein *et al.*, eds) pp. 33–36. Cambridge University Press, London.

Weiss, E. H., Mellor, A., Golden, L., Fahrner, K., Simpson, E., Hurst, J. and Flavell, R. A. (1983). The structure of a mutant H-2 gene suggests that the generation of polymorphism in H-2 genes may occur by gene conversion-like events. *Nature, Lond.* **301**, 671–674.

Weiss, E. H., Golden, L., Fahrner, K., Mellor, A. L., Devlin, J. J., Bullman, H., Tiddens, H., Bud, H. and Flavell, R. A. (1984). Organization and evolution of the class I gene family in the major histocompatibility complex of the C57BL/10 mouse. *Nature, Lond.* **310**, 650–655.

White, P. C., Chaplin, D. D., Weis, J. H., Dupont, B., New, M. I. and Seidman, J. G. (1984). Two steroid 21-hydroxylase genes are located in the murine S region. *Nature, Lond.* **312**, 465–467.

Widera, G. and Flavell, R. A. (1984). The nucleotide sequence of the murine I-E$_\beta^b$ immune response gene: evidence for gene conversion events in class II genes of the major histocompatibility complex. *EMBO J.* **3**, 1221–1225.

Widera, G. and Flavell, R. A. (1985). The I region of the C57BL/10 mouse: characterization and physical linkage to H-2K of a novel SBβ-like class II pseudogene ψAβ3. *Proc. natn. Acad. Sci. USA,* **82**, 5500–5504.

Williams, A. F. (1985). Immunoglobulin-related domains for cell surface recognition. *Nature, Lond.* **314**, 579–580.

Winoto, A., Steinmetz, M. and Hood, L. (1983). Genetic mapping in the major histocompatibility complex by restriction enzyme site polymorphisms: most mouse class I genes map to the Tla complex. *Proc. natn. Acad. Sci. USA* **80**, 3425–3429.

Winoto, A., Mjolsness, S. and Hood, L. (1985). Genomic organization of the gene encoding the mouse T-cell receptor α chain. *Nature, Lond.,* **316**, 832–836.

Wood, C. and Tonegawa, S. (1983). Diversity and joining segments of mouse immunoglobulin heavy chain genes are closely linked and in the same orientation: implications for the joining mechanism. *Proc. natn. Acad. Sci. USA* **80**, 3030–3034.

Yamamoto, T., Davis, C. G., Brown, M. S., Schneider, W. J., Casey, M. L., Goldstein, J. L. and Russell, D. W. (1984). The human LDL receptor: a cysteine-rich protein with multiple Alu sequences in its mRNA. *Cell* **39**, 27–38.

Yancopoulos, G. D., Desiderio, S. V., Paskind, M., Klerney, J. F., Baltimore, D. and Alt, F. W. (1984). Preferential utilization of the most J$_H$-proximal V$_H$ gene segments in pre-B-cell lines. *Nature, Lond.* **311**, 727–733.

Yokoyama, K., Stockert, E., Pease, L. R., Obata, Y., Old, L. J. and Nathenson, S. G. (1983). Polymorphism and diversity in the Tla gene system. *Immunogenetics* **18**, 445–451.

Yoshikai, Y., Anatoniou, D., Clark, S. P., Yanagi, Y., Sangster, R., van den Elsen, P., Terhorst, C. and Mak, T. W. (1984). Sequence and expression of transcripts of the human T-cell receptor β-chain genes. *Nature, Lond.* **312**, 521–524.